Age of System

Age of System

Understanding the Development of Modern Social Science

Hunter Heyck

Johns Hopkins University Press
Baltimore

© 2015 Johns Hopkins University Press
All rights reserved. Published 2015
Printed in the United States of America on acid-free paper
9 8 7 6 5 4 3 2 1

Johns Hopkins University Press
2715 North Charles Street
Baltimore, Maryland 21218-4363
www.press.jhu.edu

Library of Congress Cataloging-in-Publication Data

Crowther-Heyck, Hunter, 1968–
   Age of system : understanding the development of modern
social science / Hunter Heyck.
      pages cm
   Includes bibliographical references and index.
   ISBN 978-1-4214-1710-3 (hardcover : alk. paper) — ISBN
978-1-4214-1711-0 (electronic) — ISBN 1-4214-1710-3
(hardcover : alk. paper) — ISBN 1-4214-1711-1 (electronic)
1. Social sciences—United States—History—20th century.
2. Social sciences—History—20th century. 3. Social sciences—
Philosophy. 4. System theory—History. I. Title.
   H53.U5C76 2015
   300.9'04—dc23   2014036892

A catalog record for this book is available from the British
Library.

*Special discounts are available for bulk purchases of this book. For
more information, please contact Special Sales at 410-516-6936 or
specialsales@press.jhu.edu.*

Johns Hopkins University Press uses environmentally friendly
book materials, including recycled text paper that is composed
of at least 30 percent post-consumer waste, whenever possible.

This one's for Dad

Contents

Acknowledgments

Portions of the introduction and chapter 1 appeared in an earlier form in "The Organizational Revolution and the Human Sciences," *Isis* 105, no. 1 (Mar. 2014): 1–31.

An earlier version of Chapter 2 was published as "Patrons of the Revolution: Ideals and Institutions in Postwar Social Science." *Isis* 97, no. 3 (Sept. 2006): 420–46.

An earlier version of chapter 4 was published as "Producing Reason," in *Cold War Social Science: Knowledge Production, Liberal Democracy, and Human Nature*, ed. Mark Solovey and Hamilton Cravens (New York: Palgrave Macmillan, 2012), 99–116.

A version of chapter 5 was published in German as "Die Moderne in der amerikanischen Wirtschaftswissenschaft," in *Macht und Geist im Kalten Krieg*, ed. Bernd Greiner, Tim B. Müller, and Claudia Weber (Hamburg: Hamburger Edition, 2011), 159–79.

All of the above are used in this volume by the kind permission of the publishers.

As someone who often writes about patronage, I would be remiss if I did not mention that support for this project was provided by a Ryskamp Fellowship from the American Council of Learned Societies and by National Science Foundation Grant No. SES-0621232. Without this support, this project would not have been possible, materially.

I also would like to thank my colleagues, who have been invaluable sources of intellectual support, not to mention good friends, especially Piers Hale, Steven Livesey, Suzanne Moon, Pete Soppelsa, and Stephen Weldon.

My deepest thanks go to my family. I have been blessed across the generations. So, thank you Mom, thank you Debbie, and thank you Max, Piper, and Finn. It would take a book much longer than this to tell you all how much I love you.

Age of System

The Organizational Revolution and the Human Sciences

The "organism" . . . can be operationally defined as an "organization," the limits of whose field depend on the investigation. Thus it might include a single cell or the universe.

M. B. McGraw, 1940

Myrtle McGraw's extraordinarily broad definition of the organism—and of the organization—reflected the view of a rising generation within the human sciences. Still a minority in 1940, this new generation and the students they trained redefined the central concepts, methods, tools, practices, and institutional relations of postwar social science. While the movement toward a new social science took different forms in different fields, these disciplinary variations shared a common theme: the embrace of a new perspective on science and nature, one that conceived of all things in terms of organization, structure, system, function, and process.

At its apogee, between 1955 and 1970, this perspective was both more widely accepted and more precisely specified than before the war, with its exponents framing all subjects of study as complex, hierarchic systems defined more by their structures than by their components. The goal of science, in this view, was to construct formal models of system behavior, and its chief method was to develop models that would enable one complex system, such as the digital computer, to simulate the behavior of another, such as the human mind. Indeed, as Claude Lévi-Strauss put it, that which "makes social-structure studies valuable is that structures are models, the formal properties of which can be compared independently of their elements."[1]

In this book, I use the terms *high modern social science* and *high modern style* to describe this new approach to science and society when I refer to it as a constellation of ideas, practices, and social relations; I use *high modern worldview* and *bureaucratic worldview* when referring to the (sometimes explicit, sometimes implicit) basic assumptions about the structure of the world and the nature of science that underlay this new, high modern social science.

High modern style characterized the work of many elite scientists, including Herbert Simon, George Miller, Talcott Parsons, C. West Churchman, James G. Miller, David Easton, Karl Deutsch, Clyde Kluckhohn, Ralph Linton, A. F. C. Wallace, Kenneth Arrow, and Paul Samuelson, as well as crucial scholar-patrons such as Chester Barnard, Warren Weaver, J. C. R. Licklider, and Robert Taylor, just to name a few. All of these scholars held prestigious positions, published widely read books and articles, and had great influence with (or were themselves) the chief patrons of social science.

High modernism was not the property of this elite alone, however. It describes the work of a broader swath of researchers as well. I base this statement on the results of an extensive survey of the American flagship journals for anthropology, economics, political science, psychology, and sociology between 1925 and 1975.[2] This survey enabled me to identify the core characteristics of high modern social science, track its trajectory over time, and investigate its connections to the advent of new patrons. One of the key findings from this survey was that articles employing the core concepts and methods of high modern social science (system, structure, function, modeling) rose from less than 7 percent of the sample in 1930 to just over 60 percent in 1970, a more than eightfold increase in little more than a generation.

Given that 60 percent is not 100 percent, my claim is *not* that high modern social science was hegemonic or that it monopolized all social science. Rather, high modern social science was an identifiable, distinctive approach to social science, which came to define the mainstream of social science between 1955 and 1975.

A secondary argument is that this approach to social science, the set of assumptions it encompassed and the exemplary work it produced, mattered beyond the walls of academe: these were the ideas, ideals, and methods of those who advised policymakers in business and government, leading Americans into a War on Poverty at home and a war in Vietnam abroad; of those who trained new elites in new schools of business and public administration; of those who wrote the basic textbooks from which a generation learned how the economy, society, polity, and even the mind worked; and of those who wrote the position papers, books, and magazine articles that helped set the terms of public discourse in an era of mass media, think tanks, and issue networks.

Given the tight linkage among individualist public choice theories, market fundamentalism, and conservative politics today, it would be easy to assume that the politics of the high modernists, by contrast, were necessarily liberal—that this is the history of the intellectual core of left-center politics.[3] To some extent, this is true. For example, one of the underlying assumptions of high modernists was that there was a "universal man" (they included women within the category) about whom one could construct a truly universal human science. For a generation schooled by Franz Boas and Gunnar Myrdal, steeled by the struggle against Nazism, and seeking to counter communism globally, this assumption ruled out the *overt* racism and racialized analyses of group differences that had been so common in the Progressive Era up through the 1920s.[4] In the context of the 1930s to the early 1970s, to move from a biological to a social discourse about group differences (be they ethnic, racial, class, or gender differences) was almost necessarily a move to the left of center. Treating the established hierarchies of race and class and gender as products of social choices rather than biological necessities was likely to be seen as a challenge to those hierarchies. Similarly, to attribute causal agency to social forces rather than to independent individuals also generally put one somewhere to the left of the pre–New Deal center.

Although high modern ideas and frameworks were important parts of the "New Deal/Cold War liberal consensus," they were not solely the property of liberal or left-center thinkers. As Daniel Rodgers has argued so well in his brilliant *Age of Fracture*, the power of structure, system, and social influence in shaping the thoughts and actions of individuals was common sense to both center-left and center-right between the 1940s and early 1970s.[5]

If *Age of Fracture* is about the emergence of the late modern understanding of self and society as the high modern worldview cracked under the strain of the many crises of the 1970s, then this book is very much its prequel—a study of the rise of this high modern outlook from the ashes of economic depression and war to a position that was both summit and precipice. In short, the purpose of this book is to explore the rise of high modern social science and to offer an explanation for how and why this new set of ideas, practices, and relations came together the way they did, when they did.

The Organizational Revolution and High Modern Social Science

The rise of this new, high modern social science, and of the bureaucratic worldview at its intellectual core, was closely linked to the "organizational revolution" in twentieth-century American business, politics, and society.[6] The connections between this larger transformation of American society and this intellectual shift existed on three levels. First, many social scientists took the new organization and scale of modern life as their central research questions. For example, the Chicago Schools of sociology and political science in the 1920s and 1930s took their missions to be the understanding of the social and political organization of the modern city; the field of public administration came into existence as a viable subdiscipline at the same time, with its central questions all having to do with the management of large-scale organizations; and virtually every social observer commented on the social meaning of the rise of big business, the rapidly increasing division of labor, the rise of the expert, and the growth in the scale of government agencies.[7]

Second, there were both great resources and great rewards for work that fit well with the values, goals, and operational needs of the new organizers of American business and politics. As is well known, the 1920s through the 1960s was the time when the great foundations and the federal government came to play major roles in scientific research.[8] One common aspect of such institutions is that they were created and managed by people who had pioneered the organization of business and politics on new scales. The foundations they created sought, in turn, to transform benevolence into a rationally organized big business.[9] Similarly, the military research agencies were created by the people who pioneered the reorganization of American forces during World War II and the early Cold War—many of whom were the same people who had reformed private philanthropy and academic research, such as Warren Weaver, Vannevar Bush, and Rowan Gaither.[10] To them, discovery and invention were processes that could be planned and managed.

Third, the organizational revolution was enabled, though not determined, by the creation of sophisticated *control technologies*. A control technology, in my usage, is a device or formalized procedure that is used to coordinate the operations of multiple components so that they function as a single unit. Control technologies have existed for a long, long time, but the development of large-scale electric-power generation and communication systems in the late nineteenth and early twentieth centuries enabled a great leap in the power and scope of such technologies, allowing the "real-time" coordination of vast, integrated systems of production, distribution, and communication.

Such control technologies, especially but not exclusively the electrical ones, provided a set of tangible models of the intangible structures of the world. At the same time, many of these control technologies were useful not only as heuristic models of nature but also as concrete instruments for investigating, representing, or controlling it. Thus, we see some scientists using these technologies "at the bench," to investigate or represent nature, some attempting to improve them through research into their fundamental properties, some using them as heuristic models to guide their research, and still others not using these technologies in their research but still being influenced in the selection of problems, concepts, or methods by the programmatic goals of the new patrons of social science.

The organizational revolution usually is described as taking place between roughly 1870 and 1920, the fifty years *before* the period under study here. A bit of explanation about period and context thus is in order. The organizational revolution was defined by the rise of large-scale organizations and bureaucratic hierarchies in business and government, a change that was accompanied by the rise of professional managers and professional technical specialists in both arenas. These were *revolutionary* developments. As Thomas Haskell, Dorothy Ross, Ted Porter and other historians of Progressive Era social science have shown, the many changes in life associated with the organizational revolution—and its economic aspect, the second industrial revolution—did encourage a new kind of social science.[11] This new social science was increasingly professionalized, specialized, and instrumentalist; it was grounded in an awareness (even preoccupation) with interdependence, change, and subjectivity; and it aspired to new levels (and kinds) of scientific rigor.

These developments, both in society and in social science, did not stop in the 1920s. Rather, large-scale organizations continued to grow, spreading into more and more areas of the economy and public life. This is especially true of governmental organizations, which took successive leaps in scale and scope in the New Deal, World War II, and Cold War eras. It was not until the 1970s that the continuing organizational revolution entered a period of crisis and doubt. This era of organization's ascendance, rather than of its explosive rise or lumbering decline, is the context in which I place high modernism and its bureaucratic worldview.

Redefining the Subject

In the period between 1920 and 1970, virtually every field of social science reconceptualized its central object of study as a system defined and given structure by a set of processes, mechanisms, or relationships. As Timothy

Mitchell has shown, economists in the late nineteenth and early twentieth centuries did not write about or analyze "the economy." Rather, they analyzed the forces that led to equilibrium (or disequilibrium) in specific markets or industries. In the 1930s and 1940s, however, "the economy" came to mean the integrated system of exchange of a specific nation, a shift reflected (and enabled) by the creation of national income and product accounts.[12]

Similar changes can be seen in the other social sciences as well. Anthropology, by the turn of the twentieth century, had defined itself as the study of culture, but by the middle third of the century, anthropologists were redefining culture itself as a largely self-contained, relationally structured system of objects, practices, and meanings.[13] Following Émile Durkheim's famous "sociological hypothesis" that a society is an entity unto itself, not simply an aggregate of individuals, sociology likewise reformulated itself in this period as the study of social structures and functions, especially social institutions and roles. The working assumption of Talcott Parsons and his generation was that a society was a largely self-contained, hierarchically structured system of relationships.[14] This assumption was one reason for the widespread interest in social status, for example, an inherently relational concept that could be ordered but not easily quantified.

Political science likewise reconceptualized its central subject, the state. Before the twentieth century, political science tended to be either the philosophical analysis of theories of government (e.g., the study of democracy versus monarchy) or the study of the specific policies of specific rulers (debate on the corn laws). In the early twentieth century, this began to change, with studies like Arthur Bentley's *The Process of Government* (1908) exploring the structures and functions of specific states with an eye to creating a generalized science of power. By the late 1930s, the state (not just its leaders) had become an entity, a system with a specific structure defined by certain institutions, processes, and relationships (and as such could be an actor itself).[15]

It is difficult to generalize about experimental psychology until it began to unify around cognitive psychology in the late 1950s, and the interwar divisions between behaviorists, Gestalt psychologists, and Freudians were very real and very fierce. Even so, all three of these warring groups understood their subjects of study to be largely self-contained, relationally structured systems of mechanisms that governed the responses of the organism as a whole. As one psychologist put it, "The locus of the behaviors no longer is placed in the organism alone but in a field within which the organism holds status as at most only a nucleus."[16] To Freudians, these mechanisms were (metaphorically) hydraulic, with the system's goal state being an equilibrium of pressures. To behaviorists, the mechanisms were associational, with the system being defined by sets of circuit-like connections between stimuli and responses. To Gestaltists, the mechanisms acted to organize perception, and thus to organize problem-solving processes; until the arrival of the computer, however, there was no good mechanical model for such processes.[17]

Parallel redefinitions took place in the natural sciences. In biology, for example, such fundamental concepts as the ecosystem, gene, cell, organism, and species all were reconceived as structures or systems.[18] Beginning with August Kekulé and his 1865 discovery of the benzene ring, chemistry increasingly became the study of molecular structures and their relations to chemical properties.[19] The 1920s through the 1960s saw this trend extend further, as resonance and other properties of isomers (that is, of chemicals with the same components but different structures) and the structures of complex organic molecules (proteins, DNA, etc.) took the center stage of inquiry.

Physics in the early twentieth century became the study of atomic structure, with the quantum revolution carrying prewar ideas to another level of systematic abstraction and formalization.[20] In quantum analyses of atomic structure, it was the number and arrangement—the organization—of subatomic particles that really mattered, since each of the three classes of such particles is homogeneous, with every electron being just like every other electron. In fact, in quantum theory, even subatomic "particles" are redefined as patterns in fields, almost unintelligible (and certainly unmeasurable) in isolation. The interwar quantum revolution also served as grist for the philosophical mill of Alfred North Whitehead, whose exposition of quantum philosophy in *Science and the Modern World* influenced leading social scientists.[21]

While the similarities in these transformations of the central object of study across fields are striking, there were important differences as well: Gestaltists were much more likely to be exponents of a bureaucratic worldview than were Freudians or hardcore Skinnerians, for example, and no two anthropologists shared exactly the same definition of culture. In addition, while the rise of high modern social science and its connection to the organizational revolution is an important story, it is not the *only* important story for social science in this period. The rise of statistical thinking and survey methods, the dramatic changes in ideas about race and gender, the shifting concepts (and realities) related to the size and scope of government, the changing relationship between social science and the state, and the challenges posed by the Great Depression, World War II, and the Cold War all are big stories that intersect with the rise of high modern social science without being explained by it (at least, not by it alone).

Indeed, just as one could tell the story of the social sciences between 1925 and 1975 as the story of the rise and demise of the universal man, with all its attendant implications and applications to racial and gender issues, one could write an excellent book about this period that focused on the "ground level" of practice—of administrators managing organizations, social workers working cases, psychologists running experiments, sociologists and political scientists conducting polls, anthropologists doing fieldwork (perhaps in the service of the military abroad), economists gathering, making, and modeling time series data, and so on—framing one's questions about the intersection of government priorities, the needs of business, methodological challenges, social or political values, and intellectual frameworks.[22]

This book is more about words and ideas than about practices and policies, but this was a time when ideas from the social sciences shaped the thoughts and actions of executives and employees, leaders and citizens. The social sciences in this period came to occupy new positions in business, government, and public and private life more generally, helping inform and validate policy decisions both large and small, and to shape how individuals understood themselves, their families, their communities, nations, and world. In short, the ideas discussed here mattered to millions in their everyday lives, whether in the form of algorithms employed to keep shelves stocked in grocery stores or assumptions deployed in grand debates about taxes and spending or in new words used in everyday conversations, words like *stress* or *information* or *dysfunctional* families or relationships.

That these ideas mattered should come as no surprise: the nation invested billions of dollars in higher education at precisely this time, so a vast new population was exposed to courses and texts espousing new, high modern ways of understanding the world. Similarly, this was the time when the professional school of administration—either public administration or business administration—reformed, expanded, and multiplied dramatically, introducing hundreds of thousands of MBAs and MPAs to administration as a high modern, social scientific enterprise. It was also the time when such administrative elites became known as "decisionmakers," which, as chapter 4 shows, was a very new way of thinking about politics, economics, administration, and leadership.

In addition, there was a tremendous "data explosion" during this period. This explosion began with the creation of national income and products accounts during the Great Depression. (Such measures were so inadequate before the depression that policymakers had no firm idea what portion of the workforce was unemployed during that economic calamity—was it 10%, 20%, 30%, more?) It accelerated during World War II and took off in the postwar period as part of the simultaneous creation of the warfare/welfare state and expansion of corporate information systems. These developments coincided, as social scientific technicians generated the new data in the course of their work for big businesses and government agencies. Like the ideas and agencies that governed their collection, these data became touchstones for political and social debate.

These were important, indeed transformative, developments in the life of the nation, and both the data explosion and the strapping of the social sciences into the policy harness are parts of the story I tell here. Indeed, one way to understand this book is to see it as the history of the main intellectual framework that people used to make sense of this mountain of new data and to frame those new policies. For it is clear that an increasingly large number of social scientists did come to share ideas and approaches that came from the same, close-knit family. The common theme that united this family, despite its many disciplinary, institutional, and individual variations, was a high modern, bureaucratic worldview. It was the new "common sense" of the "American century."

Fittingly, this bureaucratization of the world picture took place on multiple levels. It involved a redefinition of the world as a complex, hierarchical system, of the various sciences as the study of the subunits of that system, and of human science as the study of that class of complex systems characterized by purposeful, adaptive (and hence path-dependent) behavior. The goal of this new science of complex, hierarchic, adaptive systems was the construction of formal models of system behavior, and its great challenge was to develop a rigorous formal language for the description and analysis of the behavior of such systems over time.

In the 1950s and 1960s, scientists who saw themselves as pursuing one of these systems-based "sciences of organized complexity" (to use Warren Weaver's phrase) adopted the program as one of their primary formal languages, seeing their proximate goals as the development of programs that would enable one complex system—such as the computer—to simulate the behavior of another, such as the human mind.[23] The program allowed one to describe contingent, sequential events in a precise, demanding language. It enabled the reduction of complex adaptive behavior to a set of hierarchically organized elementary processes. It permitted one to generate unpredictable complexity from determinate simplicity. It was a structure that enacted a process. Because the computer was, in a sense, the perfect bureaucracy, a machine defined almost wholly by its organization, it seemed to many to be the perfect device with which to model a bureaucratic world.[24]

The Bureaucratic Worldview and High Modern Social Science

While the advocates of the bureaucratic worldview avidly sought to understand organizations and to rationalize their structures and functions, most disliked "bureaucracy" and few were actual bureaucrats or managers (though managers in large organizations were a key part of the audience for their work). So why, then, call their worldview *bureaucratic*?

The term accurately captures their fascination with order, system, and hierarchic organization; their interest in "production systems" for knowledge, decisions, and experiments; their ties to new patrons; and their interest in the eventual practical application of research to the management of large organizations and grand technopolitical projects. Most importantly, the notion of a bureaucratic worldview at the heart of an emerging, high modern social science links a new set of ideas, assumptions, metaphors, and goals in science to its broader social context, just as the idea of a new mechanical worldview taking hold during a seventeenth-century crisis of authority located that new perspective in its time.

This bureaucratic worldview should be understood as a family of related concepts and goals rather than a catechism of required beliefs. And, like all families, it should be understood as something that changed over time. However, just as a family photo is a useful entry point into that family's history, so too a snapshot of this worldview can help us understand its history. The key members of the family are as follows:

- A prefiguration of all subjects of study as *complex, hierarchic systems*, with a correlate emphasis on the *behavioral-functional analysis* of organizational or *structural-relational properties*, especially those properties (such as *communications and control systems*) that enable internal coordination of the system so as to adapt to changing conditions and thereby to maintain *equilibrium.*

- An acceptance of some form of *analytic realism,* embracing the use of mental constructs to describe systems of often intangible but very functional relationships.[25]
- An acceptance of at least a *weak holism,* based on the belief that new properties emerge at successively higher levels in the organizational hierarchy.[26]
- An idealization of *formal, instrumental reason,* epitomized by the development of systematic *theory* and formal *modeling,* and by a fascination with *procedural logic,* as in the development of *algorithms, heuristics, protocols, decision rules, production systems,* and *programs.*
- A belief, both intellectual and moral, in a "universal man"—that is, the belief that all people have the same basic biological, mental/ psychological, and social capacities, limitations, and needs, such that a "science of man" could and should be a science of all men. This belief, as Jamie Cohen-Cole has shown, often was attached to a liberal-centrist faith in the crucial role of an "open mind" in a modern, complex democratic society, with the open mind being one that, by definition, was tolerant, reasonable, adaptable, creative, and averse to political or emotional extremes.[27]
- An abiding interest in the means by which systems *store, process, and communicate information* about themselves and their environments, often expressed through the *formal analysis of information* and *the symbols that represent it* (e.g., information theory, symbolic logic, Chomskyan linguistics, analytic philosophy).[28]

In short, the bureaucratic worldview was based on the prefiguration of the world as a set of complex, hierarchic systems (a treelike structure), with the most interesting systems having the additional characteristic of being adaptive— that is, of being environmentally responsive. Hence the proliferation of tree structures in midcentury science: organization charts (usually described as pyramids but equally well described as inverted trees), decision trees in decision theory, treelike mappings of strategies in game theory, trees in Claude Shannon's information and communication theory (in which a message is decoded through a process of selection among alternatives, with each bit of information reducing the branching of the message tree by a factor of two), linguistic trees—both diachronic, in the evolution of languages and their words, and synchronic in the mappings of the structures of languages and their grammars—semantic trees, fractal trees, genetic trees, evolutionary trees, descriptions of the nervous system as having a treelike structure, and, of course, the myriad trees in computing: file hierarchies, complete with "roots" and branches, program structures (which, fittingly, soon come to have "nests" in them, as computer scientists discover that recursion and feedback are vital complements to tree structures), and perhaps the most important tree structure in our lives today, the Internet.

Control Technologies

Having sketched the outlines of this forest of tree structures, programs, "nests" of feedback loops, and so on, we are faced with the historian's questions: the when, the where, the how, and the why. To answer these questions, we must turn to something not usually seen as important for the history of the social sciences: the history of technology, especially the history of that type of technology most characteristic of the organizational revolution, control technologies.

Indeed, the idea of control technologies is important enough to my project that it might best be seen as part of a growing body of work that treats technologies as "tools to think with"—that is, as sources of potent models and metaphors as well as new experiences and new data. Work in this tradition crosses disciplinary lines, both past and present. At its best, it brings together insights from the history of technology, the conceptual histories of various specific fields, and the history of models, metaphors, and representations in science, connecting them to the broader intellectual currents of the day while grounding them in the specifics of institutional context and instrumental practice. Some classic examples of this type of analysis include Jay Bolter's work on the crucial importance of the spinning wheel in ancient Greek and Roman thought, Otto Mayr's study of the role that automatic machinery played in the development of new ideas about liberty and authority in the early modern period, and Norton Wise and Crosbie Smith's exploration of the ramifications of the momentous shift from the balance to the engine as the fundamental metaphorical referent underlying both political economy and physics in the nineteenth century.[29]

Similarly, Anson Rabinbach has argued that the idea of the "human motor" lay at the heart of a wide variety of new ideas, attitudes, and institutions in the late nineteenth century, and Laura Otis has noted the vital role that the telegraph and telephone played as models for biological, physical, and social systems, especially the human nervous system.[30] The computer and, more recently, the Internet likewise have become central supports for many new ideas about mind, body, nature, and society, as Paul Edwards, N. Katherine Hayles, Manuel Castells, and a host of other scholars have shown.[31]

To this listing of metaphorically resonant technologies one must add organizational technologies, such as the factory and the assembly line. Like their more material counterparts, these organizational technologies were intimately connected to broader ideas about mind, body, nature, and society. Simon Schaffer, for example, has argued that Charles Babbage and his intellectual allies saw the human mind as a factory and mathematical logic as rational production system.[32] Emily Martin has found that women's bodies in the twentieth century typically were understood as factorylike production systems, particularly by physicians interested in managing female reproductive processes.[33] Similarly, Donald Worster, Peter Taylor, and Sharon Kingsland all have explored the intimate links between new ideas about economic production systems in the early twentieth century and the emergence of the idea of the ecosystem.[34]

The factory and the assembly line, however, are but two of the three master organizational technologies of the modern era. The third is the bureaucracy, which has not yet been recognized as being a fertile source of models and metaphors in science.[35] In some ways this lack of recognition is surprising, for historians and social scientists long have held that the central development of the second industrial revolution was the creation of large-scale organizations, specifically the great industrial corporations and the governmental bureaucracies that regulate them. Indeed, from Max Weber to Talcott Parsons to Alfred Chandler to Robert Wiebe and the pioneers of the "organizational synthesis" of American history, bureaucracy has been seen to be one of the defining social forms of the twentieth century—for good or for ill.

In other ways, however, the relative lack of attention to the conceptual significance of bureaucracy is not surprising. Bureaucracies are intangible things, after all, despite the myriad monuments of glass and steel that have been built to house them. As a result, one might expect that bureaucracy would be a less vivid metaphorical referent than the potter's wheel, the clock, or the computer. In addition, in all the studies mentioned above, the technologies in question acquired the ability to serve as heuristic models for thinking in a wide variety of fields because they changed understandings of the human body, which was and is a basic metaphorical referent for human thought, as George Lakoff and

Mark Johnson have argued.[36] Humans may not be the measure of all things, but they are the ones who measure and thus the ones who count. Until one could build a connection between the bureaucracy and the body, then, bureaucracy likely would be only a weak support for metaphors and analogies.

The first step in linking the bureaucracy and the body was the idea that the organism is an organization, and vice-versa, an idea new to Romantic biology that flourished in the nineteenth and early twentieth century, especially in the new physiology of Claude Bernard and Herman von Helmholtz.[37] The next, crucial step came with the development and widespread use of new control technologies.

Recall that the goal of a control technology is to enable dispersed, heterogeneous components to function as a single system. Another way of putting it would be to borrow Robert Wiebe's famous phrase from *The Search for Order* and say that control technologies are made to facilitate "continuous management."[38] Some examples of such technologies would be governors and other feedback devices, information technologies (such as the telegraph, telephone, digital computer, and computer networks), organizational technologies (formalized procedures for measuring, monitoring, scheduling, and processing, of which the assembly line is the great exemplar), and, with the arrival of electricity, power transmission technologies. Electrification plays a crucial role in this history, for electricity, remarkably, can carry both information and power, can be used to connect multiple devices, and can move so fast that continuous management can become continuous, *real-time* management. It also loves being in closed systems and so encourages the construction of such structures.

Control technologies are thus the essential other side of the coin of the division of labor and specialization of function that is so characteristic of modern organizations and of modern life more generally. It is no accident that both the division of labor *and* coordination, order, and structure were fascinating to so many in the human sciences in the twentieth century.[39] It is these new control technologies that provided the essential material and intellectual spark for the creation of the bureaucratic worldview—a connection that some of its promoters read back into its past. Herbert Simon once wrote, "Physicists and electrical engineers had little to do with the invention of the digital computer . . . the real inventor was the economist Adam Smith, whose idea was translated into hardware through successive stages of development by two mathematicians, Prony and Babbage."[40]

I should note, however, that I do not intend to paint a deterministic picture, wherein the creation of new control technologies *necessarily* led to this new view. People certainly could have thought and acted differently, and many did. Rather, I believe that sophisticated control technologies, especially electrical ones, provided a set of tangible models of the intangible structures of the world for the many who were looking for just such models. At the same time, the promise such technologies held for economic (and later, military) reward meant that there was rich soil in which such approaches could take root.

The Payoff: A New Frame

In this book I argue for the existence of a new, high modern social science; characterize this high modern social science as a particular instantiation of a new outlook, the bureaucratic, high modern worldview; and argue that the emergence, rise, and eventual fragmentation of this high modern style is best understood in the context of the continuing organizational revolution, exploring connections between this context and high modern social science in concepts and problem choices, in institutional relations (including patronage relations), and in the crucial role of control technologies as tools, models, and metaphors. This framing thus links the history of the social sciences to the history of technology and the history of science, and it connects all three to broader narratives of modern American history.

I have no illusions that this approach is the "one best" way to understand these different histories, and the web of connections among them can be traced in many ways. This way of framing things, however, does have the potential to help us shed new light on each of these histories.

For example, taking this approach to the history of the postwar social sciences helps clarify the ways they were, and were not, products of the Cold War. The Cold War clearly is a crucial part of the active broader context for the rise and fracture of high modern social science, and sometimes it moves from background to center stage: the institutional transformation described in chapter 2 is part of the Cold War transformation of scientific patronage; modernization theory (chapter 5) obviously is wrapped up in Cold War politics (Walt Rostow's paradigmatic work, *The Stages of Economic Growth*, is subtitled *A Non-Communist Manifesto*); many of rational choice theory's first great exponents worked at RAND, the quintessential Cold War think tank (discussed in chapter 4); the new model of man described in chapter 3 almost could be called "Cold War liberal consensus man," for it projected the model citizen for a regulated market and rationalized democracy, as Jamie Cohen-Cole argues in his incisive new book, *The Open Mind*; and modeling (chapter 6) certainly fits the simultaneous flight to abstraction of many Cold War sciences eager for government money (and therefore wary of "politics"—as George Reisch argues regarding philosophy in *To the Icy Slopes of Logic*) and the stampede to application of others eager to prove their fields' practical worth to themselves and their government patrons.[41]

At the same time, *Age of System*'s core argument about the relationship of this high modern style to the ongoing organizational revolution does not *depend on* the Cold War. It *does* depend on industrial modernity, which in the United States took the form it did in part because of the contingencies of the Great Depression, World War II, and the Cold War.[42] Thus, it tells us something about modernity and how those scholars who took on the task of understanding—and therefore shaping—modernity approached that task in twentieth-century America.

In order to do so, I begin with a broad overview of high modern social science's intellectual and institutional characteristics. Chapter 1 gives a bird's-eye view of high modern social science and its bureaucratic worldview, tracing the emergence, redefinition, and spread of certain key concepts between the 1920s and 1970s. Chapter 2 looks at high modern social science's characteristic institutional forms and patronage relations. Chapter 3 takes a "horizontal" slice of the high modern social sciences in 1956, illuminating the relationships among key ideas, individuals, and institutions in the year of high modernism's "takeoff."

Having established the broad shape of the branching tree of high modern social science, I explore some of its most significant intellectual fruits: modernization theory, choice theory, and modeling. Chapters 4–6 trace the development of these fields and approaches as they developed over time, with chapter 4 exploring ideas about choice and reason from the end of the war through the 1970s, chapter 5 looking at ideas about modernity and modernization from the interwar period to the present, and chapter 6 examining the rise of models and modeling from the 1950s to the present.

The book concludes with a look at the legacies of high modern social science. Here, the value of linking that science to the continuing organizational revolution and its associated techno-social projects, problems, and control technologies becomes even clearer, for this link helps us understand some of the changes—and crises—that have beset so many fields since the 1970s. The increased interest since the mid-1970s in networks (rather than systems), chaos and complexity (as opposed to organization and hierarchy), flexibility and the spontaneous production of order from disorder (as opposed to stability produced through continuous management), and contextual, situated knowledge (as opposed to formal, instrumental knowledge) can be seen as a reaction against the narrower forms of the bureaucratic worldview, a reaction concurrent with broader public disenchantment with bureaucratic rigidities.

A new ideology has emerged that takes choice, not system, as its cornerstone, and its adherents dwell uneasily amid a vast inheritance of techno-social infrastructures built on the logic of system, structure, and control. The adherents of the bureaucratic worldview marched under the sign of the tree; today the net seems a more congenial symbol. Yet, those who embrace the net would do well to remember that today's Web was woven amid the branches of the tree of organization.

CHAPTER ONE

High Modern Social Science
A Bird's-Eye View

For political science, that is, the individual is a fiction.

Arthur Bentley, 1949

Before 1940, *zero* research articles published in the flagship journals of the five largest social sciences in America described what they were doing as "modeling" something.[1] Zero. By the 1970s, *half* of all articles in those journals did so. Before 1950, only a tiny handful of articles employed any mathematics more complicated than a simple count; by the 1970s, 60 percent of them employed mathematics that went beyond counts and measures. Before 1950, it was rare for an article to cite any form of external support for research; the three most influential patrons of social science for the next twenty-five years (the Ford Foundation, National Science Foundation, and National Institutes for Health) did not yet exist. By the 1970s, it was rare for an article *not* to cite a patron, and the "big three" loomed large.

Before 1950, few articles in these journals connected their empirical findings or philosophical discourses to anything like what the coming generation would call a theory; by the 1970s, it was extraordinary for an article *not* to frame its discussion of empirical or experimental particulars in terms of their relevance to a particular theory or model. Before 1950, talk of "systems" or "structures" was uncommon and very loose, usually implying little more than the basic presupposition that societies, economies, polities, cultures, and individual humans were organized things and thus were amenable to scientific analysis. By the 1960s, *system* and *structure* and associated words, such as *function* and *process*, were widely used terms carrying specific conceptual freight.

Conversely, before 1950, roughly 20 percent of all research articles directly addressed current social issues or public policy questions, with many more pages in the *American Economic Review*, the *American Journal of Sociology*, and the *American Political Science Review* being devoted to other kinds of pieces on the issues of the day. In the postwar era, however, fewer than 5 percent of all research articles directly addressed such questions, and the other writing on current issues nearly vanished.

At the same time, while it was quite uncommon in the 1920s and 1930s for a research article to explain group differences as having a basis in biology (e.g., articles advocating eugenic policies or describing racial differences on intelligence tests), such articles did exist: 2.2 percent, with most of those coming in 1925 and 1930. After 1945, however, such pieces were rare in the extreme (only 0.5%, or 6 out of 1,107 articles), with almost all such articles being relatively racially neutral (e.g., a 1975 article finding a genetic link to lactase deficiency, as compared to a 1930 article advocating "eugenic" sterilization of the "unfit."). Indeed, articles explicitly condemning the "old," "outdated," "tired," "racial dogmas" begin to appear as early as 1925 and from 1935 on outnumbered such eugenic or sociobiological articles roughly five to one.

In short, if the articles in these flagship journals are any guide, in little more than a generation there had been a sea change in what social scientists thought was their basic intellectual task. The ideal product of social scientific research no longer was an erudite commentary on the state of modern man, one that might educate the reader morally or philosophically or inform him scientifically about the grand issues of the day. Nor was it a description of empirical observations, untethered to theory.

Rather, the ideal product was a model of the structure of some kind of system, with everything from individual organisms to businesses to nations being understood as systems. As one psychologist wrote, "The locus of the behaviors no longer is placed in the organism alone but in a field within which the organism holds status as at most only a nucleus."[2] In short, the subjects of social science were systems structured by relations, the method employed was behavioral-functional analysis, and the goal was a theoretical model, one that potentially could be made an operational guide to practical action—in some other, future publication.

This chapter charts the contours of the emergence, development, and spread of this new, high modern social science through a survey of research articles published in the American flagship journals for economics, sociology, political science, psychology, and anthropology between 1925 and 1975. Practitioners of this high modern social science shared a bureaucratic worldview, not an ideology per se but a range of "familial" variations on common conceptual themes, as can be seen in the work of scholar-patrons Chester Barnard, Warren Weaver, and the team of J. C. R. Licklider and Robert Taylor. In their work, the similarities between humans and machines moved from being considered as metaphors to being formal concepts and design goals. These men and their allies also helped shape new patronage systems that both supported and shaped social science research (chapter 2). By 1956, the annus mirabilis of high modern social science, these ideas—and their support networks—were in full flower, as a new model of humans, science, and society took root (chapter 3).

The Survey

It is widely known that there was a behavioral revolution in political science, a cognitive revolution in psychology, a structural-functional revolution in sociology, a structuralist revolution in anthropology, and macroeconomic, econometric, and game-theoretic revolutions in economics during the 1950s and 1960s.[3] It is also known that all of these revolutions involved rebellions against traditional moral-philosophical approaches to social science. What is not well known is whether and how these revolutions within the disciplines were related to one another and to broader changes in American intellectual and political culture.

My survey of the research articles shows that these multiple revolutions had much more in common with each other than a basic behavioralist positivism or simple scientism. Rather, this survey reveals an identifiable set of philosophical, conceptual, methodological, and institutional commitments that were widely shared across fields, with the specific forms those commitments took differing from field to field. In short, what we see is a cross-disciplinary theme with disciplinary variations.

Perhaps the best evidence of this cross-disciplinary theme is that significant numbers of articles exemplifying high modern social science came from all five disciplines: *American Anthropologist, American Economic Review, American Journal of Sociology (AJS),* and *Psychological Review (PR)* each contributed between 18 and 24 percent of sampled articles that can clearly be identified as members of the family of high modern social science. The *American Political Science Review (APSR)* is the only one contributing markedly fewer to the total (a little over 14%).[4] This smaller contribution, however, is due in large part to the significantly smaller number of articles published in the *APSR* overall (about a hundred less than the average of the other four). If it had published the same number of articles as the journal with the next fewest, the *PR*, then it probably would have contributed around 16 percent (and the others 17–23%); it would still be something of an outlier, but a much smaller one.[5]

This picture of a common theme with local variations indicates that at least two levels of explanation and contextualization are needed: one that comprises all the disciplines and explains the common features of this movement across fields and one that explains the variations on this theme that took place within each field.

The High Modern Worldview

The common conceptual theme in high modern social science was the embrace of the bureaucratic worldview (see the introduction for a list of its key components). In short, this worldview was based on the prefiguration of the world as a nested set of complex, hierarchic systems (a treelike structure), with the most interesting systems having the additional characteristic of being adaptive—that is, of being environmentally responsive.

These commitments and interests certainly characterized the work of many leading social scientists, such as Herbert Simon, Talcott Parsons, James G. Miller, David Easton, Clyde Kluckhohn, and Paul Samuelson, and they could be seen in the guiding principles of leading patrons.[6] But were these views the property of a select elite, or did they characterize a broader swath of researchers?

The survey confirms the rise of a new, high modern social science in the postwar period, with high modern concepts and methods characterizing a majority of the articles published between 1960 and 1975 in the anthropology, political science, psychology, and sociology journals and a plurality of those in the *American Economic Review*.

The central concepts of this approach were *system*, *structure*, and *function*, and the ideal product (not always realized in practice) was a *model* of a system's structures, functions, and relations, preferably one based on quantitative behavioral data. Articles tagged with at least one of these four keywords rose from a low of 7 percent of all articles in 1930 to a full 60 percent of all articles in 1970, a more than eightfold increase, before declining to 52 percent in 1975 (figure 1.1).[7]

Even at its peak, however, high modern social science was not hegemonic. It clearly was the mainstream from 1955, being far more common than any single competing approach, but other approaches together constituted either a majority or a large minority at all times. For example, while some versions and aspects of psychoanalytic theory fit well with a bureaucratic worldview,

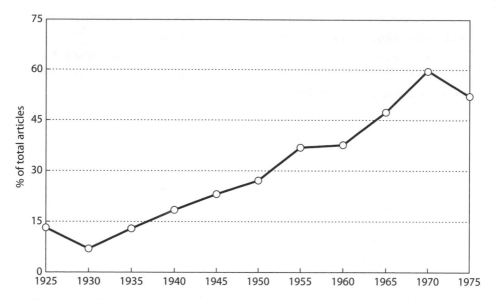

Figure 1.1. Articles in the Outer Core. Includes those tagged with at least one of four main keywords, minus those for which *model* was the only keyword used out of the top twenty-five.

most high modernists found psychoanalysis far too speculative and unscientific. Likewise, the rapid rise of survey research and statistical analyses were partially independent events: many pollsters and statisticians were high modernists, but some were not. And there were many specific topics and issues of great importance—race, in particular—that had no *necessary* connection to the bureaucratic worldview, though high modernists did interpret race through the filters of system, structure, function, and process, and their belief in a universal science of "man" ruled out certain older forms of racial (or at least bio-social) thinking.

The term *high modern social science* refers to a constellation of ideas, practices, and social relations; *bureaucratic worldview* describes the (sometimes explicit, sometimes implicit) basic assumptions about the structure of the world and the nature of science that underlay high modern social science.

Distinguishing between the worldview underlying high modern social science and the actual instantiation of that worldview in practice is important because there are always multiple ways for underlying ideas to play out in specific situations. For example, a belief in complexity and interdependence in nature need not lead to an emphasis on the value of interdisciplinary work (though those ideas "fit" well together), and even if it did, it would not necessarily lead to an emphasis on conducting interdisciplinary work in teams, as opposed to training people to be interdisciplinary generalists, say. Similarly, assuming that the social world is a system need not lead to an interest in communications in that system; one could be fascinated by energy flows (among many other things) instead.

Thus, *high modern social science* refers to the specific set of ideas, practices, and social relations, associated with the bureaucratic worldview, that was actually instantiated in the period under study. If one were to give it an operational definition, probably the best would be: the set of articles, authors, ideas, institutions, patrons, and practices found within the limits of the "Outer Core" category of articles in the survey. (See appendix for a detailed discussion of these categories, survey methods, and findings.)

High modern social science is a specific form of modernist social science: "high modern" because *high* evokes both the lofty ambition and the sometimes overweening confidence that high modernists possessed. It also reflects the universalizing "view from above" that characterized much analysis.[8] In addition, it leads to calling the social science of the 1970s and 1980s "late modern," which captures both its continuing modernist heritage and the widely felt sense of impending crisis of that period.[9]

Detailed Results

Using the four main keywords (*system, structure, function, model*) and twenty-one secondary keywords related to high modern social science, I divided the total universe of articles surveyed into the following concentric circles (each of which includes those within):

- The innermost circle of *Epitomes*, which includes articles tagged with three or more of the four main keywords ($N = 50$).
- A larger *Inner Core*, which includes articles tagged with two or more of the four main keywords ($N = 226$).
- A *Core*, which includes articles tagged with at least one of the four main keywords and one or more of the 21 secondary keywords ($N = 444$).
- An *Outer Core*, which includes articles tagged with at least one of the four main keywords, minus articles that used the term *model* without using any other of the main or secondary keywords ($N = 597$).

- A yet wider circle of *Affiliates*, which includes articles tagged with at least one of the main *or* secondary keywords, minus articles that use the term *model* without using any other relevant keyword ($N = 849$).
- The widest circle of articles that might be considered examples of high modern social science, the *Margins*, which includes articles tagged with at least one of the main or secondary keywords, with no subtractions ($N = 924$).
- The remainder, the set of *Outsiders*, which includes articles not tagged with any of the main or secondary keywords ($N = 904$).[10]

This division into concentric circles fell out rather easily from the survey data, with articles in the Inner Core immediately giving the reader the impression that they were "doing the same thing" in one field that other articles were doing in other fields (e.g., articles in the *APSR* applying Parsonian theory to the political system or ones in the *AJS* applying structural anthropology to the study of American society). Those in the Core and Outer Core also shared obvious similarities, with articles in the Outer Core being much more like those in the Core than those in the Affiliates category. Articles in the Affiliates group shared important features with ones from the Core, but they were noticeably more diverse in approach, with systems-oriented, structuralist, or functionalist analysis often being one of several modes employed (figures 1.2–1.3).

Each of the four main keywords was used as a label for over 200 articles, markedly more than any other single keyword, except one. That one was *behavior* (as in, observed behavior is taken to be the proper subject for social science), which also was used as a label for over 200 articles. For reasons discussed below, however, behavioralism was best treated as something strongly correlated with high modern social science rather than as a defining feature of it.

The next most commonly used keywords were *process* (in the sense of a law-governed mechanism operating over time, in a defined sequence) and *hierarchy* (used to code articles in which the hierarchic nature of the object of study was an important aspect of the analysis). Reconciling process and hierarchy was difficult but not absolutely impossible: nine articles were labeled with both terms.

The only major divergence from the common pattern in the four chief keywords is that while the other keywords see a rapid rise beginning in 1950 and an incipient decline in 1975, *model* continues to rise unabated, a rise linked to its becoming a near-universal aspect of economics and a very common term

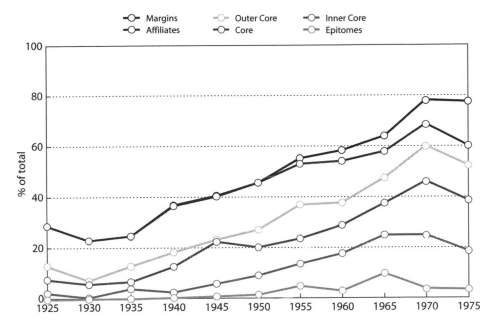

Figure 1.2. Articles in the six circles. The Margins include articles tagged with at least one of the top twenty-five keywords. The Affiliates, at least one of the top twenty-five keywords, minus those for which *model* is the only keyword used of the top twenty-five. The Outer Core, at least one of four main keywords, minus *model* only. The Core, at least one of four main keywords, minus *model* only, and two or more of the top twenty-five keywords. The Inner Core, two or more of the four main keywords. The Epitomes, three or more of the four main keywords.

in cognitive psychology, two trends that overwhelm a slight decline in usage in the other social sciences during the 1970s. Modeling resumed its upward climb in other fields by the 1980s even as its cemented its central role in economics and psychology (see chapter 6). Thus, modeling in social science had two related, but distinct, careers: a first one as part of high modern social science and a second one in which it was no longer the universal method sought by high modernists but a whole universe of methods embraced by social scientists across a range of fields.

Correlations and Connections

After charting the incidence of articles tagged with the keywords discussed above, the next step was to attempt to correlate articles exemplifying high modern social science with other factors and features. Specifically, the high

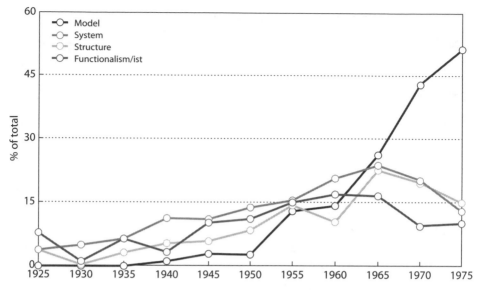

Figure 1.3. Articles tagged with the four main keywords.

modern social science sets, from Epitomes to Affiliates, were compared to the sets of articles expressing the following: a strong commitment to theory (the *theory* set); explicit commitments to behavioralism or behaviorism (the *behavior* set); a commitment to some form of dynamic analysis or analysis of change over time (the *dynamics* set); a strong commitment to mathematization and quantification (the *math* set); a hierarchical understanding of the world's structure (the *hierarchy* set); a strong interest in current public issues (the *public policy* set); a strong interest in communication, language, representation, and symbols (the *communication* set); a strong interest in coordination, integration, and other mechanisms of social control (the *control* set); or a commitment to a more traditional moral-philosophical approach to social science (the *moral philosophy* set). Additionally, they were compared to the sets of articles that cite a source of patronage (the *patronage* set), list multiple researchers or authors (the *team* set); or ground an analysis of group characteristics in biology or heredity (the *sociobiological* set).

If my thesis about the nature of the bureaucratic worldview is correct, then articles exemplifying high modern social science should be strongly positively correlated with the *theory, math, behavior, communication, hierarchy, control, patronage,* and *team* sets, all of which should have grown markedly after World

War II. The notion of a bureaucratic worldview underlying high modern social science did not lead to any obvious predictions about trends in analyses of public policy, though it did suggest that there would be a negative correlation between articles exemplifying high modern social science and ones embracing a more traditional moral-philosophical approach and those adhering to socio-biological or hereditarian approaches. All of these predictions were borne out, though sometimes in surprising ways.

Two basic methods were used to test these correlations. The first was to track the overlap between each set and each circle, from the Epitomes to the Margins, to see if the overlap grew stronger or weaker as one moved toward the heart of high modern social science (table 1.1). The second was to compare the percentage of articles in the Core that overlapped with the relevant set to the percentage of articles in the Outsider category that overlapped with the relevant set (table 1.2). The Core was used for comparison because it was the largest set whose members were nearly unquestionable exemplars of high modern social science, though, as it turns out, the ratios change but little between the Outer Core and the Core. The percentage overlaps were then compared to each other, forming a simple ratio indicating how much more (or less) likely an article in the Core was to be part of the relevant set than was an Outsider article.

This second measure was then supplemented by calculation of an "odds ratio"—a simple, commonly used statistical test of the strength of the association between two variables: for example, the odds that an item in the Core was also a member of the hierarchy set are nearly three times the odds that an item in the Outsider category was. One also can explore the relationship by going the other way, dividing, say, the odds an item in the theory set had of being in the Core by the odds of it being in the Outsiders category. In this case, articles in the theory set were eleven times more likely to be in the Core than in the Outsiders. Numbers less than 1 reflect a negative correlation: for example, the odds of an article in the public policy set being in the Outsider set are roughly 100 times the odds that it will be in the Core.

Briefly, these comparisons reveal that high modern social science was *very* strongly correlated with an emphasis on theory, with articles in the Core being up to twenty times as likely as Outsiders to emphasize theory.[11] Similarly, there were strong correlations with behavioralism, with an interest in language and communications, in social coordination and control mechanisms, in change and dynamic theory, and in organizational properties generally and hierarchic

Table 1.1 Percentage of overlap between articles in various circles and sets

Set	Outsiders	Margins	Affiliates	Outer Core	Core	Inner Core	Epitomes	Track
Theory	3.9	36.0	31.3	42.0	44.4	51.8	64.0	Inward
Communication	1.5	8.8	9.5	11.4	13.5	16.8	30.0	Inward
Behavior	17.0	54.0	50.0	60.0	64.0	70.0	84.0	Inward
Math	17.6	49.7	45.2	57.3	59.7	67.3	74.0	Inward
Dynamics	8.0	32.0	33.0	32.0	37.0	37.0	34.0	Weak inward
Control	4.9	25.0	27.2	20.4	24.8	20.4	24.0	Flat
Theory (minus model)	3.9	12.6	14.0	15.4	16.7	21.2	40.0	Inward
Organization	7.0	19.0	21.0	21.3	22.3	19.0	22.0	Flat
Patronage	15.5	33.7	33.1	38.2	38.1	41.6	38.0	Weak inward
Hierarchy	10.0	21.2	22.9	20.4	23.6	23.9	28.0	Weak inward
Team	14.8	19.9	19.0	19.8	20.9	22.1	20.0	Flat
Moral philosophy	15.2	9.3	10.1	10.1	8.6	8.8	4.0	Weak outward
Public policy	21.0	6.1	5.3	2.8	1.8	0.0	0.0	Outward

Table 1.2 Percentage and odds ratios, Core versus Outsiders

Set	Percentage Ratio*	Odds Ratio 1**	Odds Ratio 2***
Theory	11.46	19.80	11.00
Communication	8.73	9.93	9.90
Behavior	3.74	8.66	2.50
Math	3.39	6.94	2.20
Dynamics	4.51	6.57	3.30
Control	5.09	6.44	3.50
Theory (minus model)	4.30	4.97	10.98
Organization	3.05	3.64	1.80
Patronage	2.46	3.35	1.30
Hierarchy	2.36	2.78	1.20
Team	1.41	1.52	0.50
Moral philosophy	0.56	0.52	0.13
Public policy	0.09	0.07	0.01

Notes: *Percent in Core also in relevant set divided by percent in Outsiders also in relevant set. **Odds an item in Core is also in relevant set divided by odds an item in Outsiders is also in relevant set. ***Odds an item in the relevant set is also in the Core divided by odds an item is also in Outsiders.

organization and social stratification in particular. There were strong correlations with the use of mathematics and with the receipt of patronage as well. There was a small positive correlation with the conduct of work in (small) teams. There was a strong negative correlation with traditional moral-philosophical approaches and very, very strong negative correlation with discussion of public policy and current social issues. The numbers for the hereditarian/bio-social set of articles were so small as to make the statistical comparison difficult, but there was essentially zero overlap between such articles and articles in the Outer Core.[12] These ratios all fit predictions, though the trend lines across the circles for the control, organization, and team research sets were fairly flat, suggesting that they were not as central to high modern social science as were commitments to theory, behavioralism, and mathematics.

Surprisingly, however, the absolute numbers for some of these overlap sets were smaller than expected. In particular, only 95 articles fit the criteria for the communications set, compared to over 350 in the hierarchy set, 275 in the control set, and 370 in the dynamics set. This small number for the communications set is hard to reconcile with the abundant evidence of widespread interest in language and communications during this time; this is the era when departments of communications or mass communication studies were first created, when linguistics acquired new intellectual power and prominence, and when a host of cybernetic and information-theoretic concepts spread throughout the social sciences.[13] As this period saw an explosion in more specialized journals and technical reports, it seems likely that such work appeared in these new venues instead of the more traditional flagship journals.

Also, the correlation with team research was not as strong as expected. Many leading social scientists in the 1950s and 1960s expressed a strong commitment to work on interdisciplinary teams, but less than 20 percent of all articles had more than one author or referenced an additional researcher as being part of the project leading to the publication.. The correlation between team research and an article being in the Core is positive, but it is due almost entirely to the very strong correlation between team research and receipt of patronage (an article is more than five times as likely to be in the team research set if it received patronage than if it did not), and the receipt of patronage was strongly correlated with other aspects of high modern social science. This dependence on the intervening patronage variable does not make the correlation between team research and high modern social science meaningless—the patrons who funded the team work funded it because it fit their conceptions of good science more broadly, not just because it was done by a team—but it does mean that support from patrons was crucial for turning the ideal of team research into a reality.

Also, while one could read an implicit acceptance of analytic realism or of weak holism into many articles in the Core, few articles engaged in any kind of explicit (or even semi-explicit) discussion of such philosophical orientations. Apparently, social scientists believed that one should not discuss one's ontology in print. As a result, while this study did not produce any evidence disproving the idea that such philosophical commitments were part of the pre-theoretical grounding of the bureaucratic worldview or of the practice of high modern social science, it did not generate much direct evidence confirming that idea either.

In addition, the relationships to behavioralism, dynamics, and interest in hierarchy or social stratification were more complicated than the ratios might indicate. Explicit behavioralism, for example, was much stronger in articles exemplifying high modern social science than in Outsider pieces, as noted above, but its historical trajectory is subtly different than for the other main keywords. Explicit endorsements of behavioralism could be found in reasonable numbers before World War II (figure 1.4). Such articles were decidedly in the minority—about 20 percent of articles—but that percentage is several times

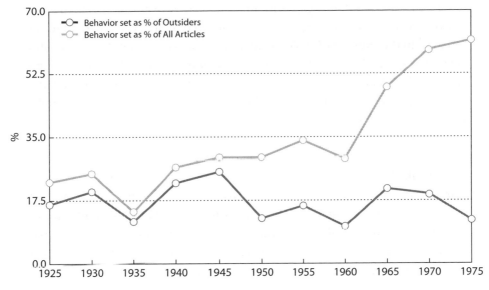

Figure 1.4. Articles in the behavior set.

higher than the percentages for the four main keywords or for other affiliated keywords at that time. Between 1950 and 1965, behavioralism and high modern social science tracked closely together. At the other end of the time scale, though, a persistent minority of Outsider articles (15–17%) continued to embrace behavioralist approaches without embracing other aspects of the bureaucratic worldview. Thus, while the odds ratio of behavioralism in the Core to that in the Outsider set overall is very high (8.66 to 1), the absolute numbers of behavioralist Outsiders remained substantial, as shown by the still positive, but markedly lower, second odds ratio reflecting the odds of a behavioralist piece being in the Core versus being an Outsider (2.5 to 1).

This pattern suggests that early behavioralism was at least partially independent of the new approach but that high modern social science and behavioralism grew to be close allies for the first twenty years after the war. This closer relationship in the postwar period was never complete unity, however, with a persistent minority of behaviorists adhering to other perspectives. In short, while the behavioral revolution and the rise of high modern social science were closely linked developments that overlapped significantly in concepts, people, and institutions, they were not synonymous.

The correlation with interest in some form of dynamic analysis (in this study, any analysis that explores meaningful change over time, from an analysis of process, sequence, or path dependence to that of development or social change, to the embrace of evolutionary models) is also curious. While the ratio of articles in the Core to Outsider articles that embraced some form of dynamics is quite high (odds ratio 6.6 to 1), interest in dynamic analysis increases only the tiniest bit as one moves in from the Affiliates to the Inner Core, a pattern unlike that for other strong correlates. The reason becomes clearer if one breaks apart the different elements of the dynamics set. Then one finds that articles emphasizing *process* rise from 16.5 percent of the Outer Core to 23 percent of the Inner Core while the other terms in the dynamics set do not increase in frequency (or dip very slightly) as one moves inward. This trend indicates that process-related concepts were more closely related to the conceptual heart of high modern social science than were other ways of approaching change over time. *Process* (used to label 176 articles overall) was a much more commonly merited label than *development* (56 articles) or *evolution* (38 articles), for instance.

The correlation with interest in hierarchy or stratification in social structures follows a similar pattern to that of behavioralism. One did not have to be an exponent of high modern social science to be interested in social stratification, status, class, or caste, which were frequent subjects of anthropological and sociological articles in the 1920s, 1930s, and 1940s, though such discussions almost always were purely descriptive until the 1940s.[14] Practitioners of high modern social science were interested in such hierarchic orders even more often, however, and they tended to assume their existence even when they were not investigating them directly. Most importantly, when they did investigate such hierarchic structures, they strove to do so via systematic theory.

Some of the most interesting results of this survey are related to patronage. They are discussed in detail in the next chapter as part of a broader consideration of the role of patronage. Here, suffice it to say that patrons supported high modern social science *preferentially*, but *not exclusively*. Other approaches to social science could find support from patrons, especially if those approaches were methodologically rigorous. The most diametrically opposed approaches (moral-philosophical and public-policy-oriented) received little funding, with public policy work receiving almost no funding whatsoever (figures 1.5–1.7).

Keywords

The journal survey used over 150 keywords to label and categorize 1,828 articles, but a much smaller, interrelated set could be used to identify exemplars

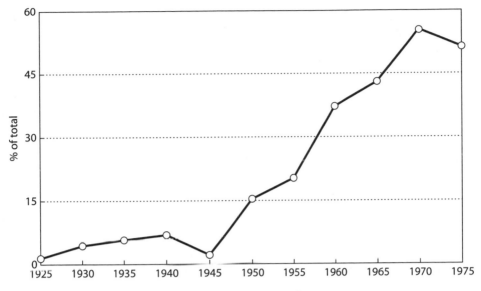

Figure 1.5. Articles citing support by patrons.

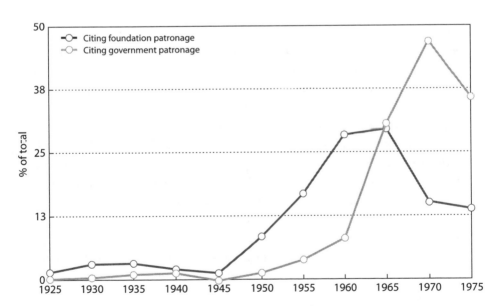

Figure 1.6. Articles citing support by foundations or civilian government patrons (mainly NIH and NSF).

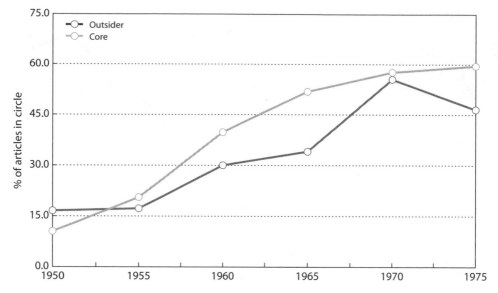

Figure 1.7. Articles citing support by patrons, Outsider and Core circles compared.

of high modern social science. The prime four keywords were *system*, *structure*, *function*, and *model*(ing), with *behavior*, *hierarchy*, *process*, *symbols/communication*, and *control* being important secondary terms. A final term that was not used frequently in the sample but that other sources have convinced me was an important, if often unstated, goal was *synthesis*.

Of these keywords, probably the most important was *system*, for to those who saw the world as a complex, hierarchic system, much of the rest followed. In this view, the economy, the family, the individual organism, the cell, the atom all were complex, hierarchically structured systems. That they were systems meant that their component elements were strongly interdependent. That they were hierarchical meant that they had a treelike structure and so were decomposable into subsystems, sub-subsystems, and so on. That they were complex meant that the behavior of the system at one level of the hierarchy was difficult to predict from knowledge of the properties of the elements at lower levels.

This prefiguration of the world as system had three main implications for both the questions high modernists asked and the methods they used to answer them. First, seeing the world as system focused their attention on systemic properties, such as the organization of a system's components, the means by which they communicated with each other, the ways that the system maintained equilibrium, and how the overall system adapted to its environment. Second, seeing the world as system encouraged a behavioral-functional mode of analysis. In this view, an organism could be known only by its behaviors, and its behaviors could be known and identified only by their effects on the other elements of the system to which the individual belonged. This was true for objects just as it was for humans: to Herbert Simon, for example, even such a seemingly natural and individual quality of an object as its mass was a property of that object in a particular system, not of the object itself.[15]

Hence, one could describe the bureaucratic worldview as the set of assumptions on which behavioralism and functionalism rested. Talcott Parsons is a perfect example: *The Structure of Social Action* (1937) and *The Social System* (1951) both defend analytic realism; emphasize system, hierarchy, and structural relations; and explore structures as they are expressed in social processes. The goal is the continuous management of society to maintain social equilibrium.

Third, seeing the world as a system encouraged mathematical formalization. To the high modernists, there was no question that a reformed behavioral science would be mathematical, for mathematics was the essential "language of discovery" of science. Paul Samuelson, for example, helped create macroeconomics (in contrast to microeconomics) as a special subject on the basis of a difference in scale and complexity, citing Alfred North Whitehead on the topic; he saw the economy as a thermodynamic system that was to be managed by a set of mathematically sophisticated economic experts in an active federal government or in large firms. In addition, Samuelson described his greatest teacher at Harvard as being E. B. Wilson, the statistician, and he dedicated his *Foundations of Economic Analysis* to J. Willard Gibbs, the mathematical physicist renowned for his codification of the mathematics of thermodynamic systems.[16]

The second great keyword is *structure*. To those who shared the bureaucratic worldview, not only is the world structured but also the object of science is to discover and model those structures. The prevalence of the idea of structure reveals connections to the structuralist movement in continental science and philosophy. In his 1970 book *Structuralism*, for example, Jean Piaget states that Gestalt psychology, Chomskyan linguistics, relativity physics, cybernetics, decision theory, group theory in mathematics (and its descendants in Kurt Gödel, David Hilbert, and Alan Turing), and his own developmental psychology all were manifestations of structuralism, which he defines as seeing things as "self-regulating systems of transformations."[17]

A computer program fits Piaget's definition of a structure almost perfectly: it is a self-contained set of symbols representing both data and instructions, both elements and rules of combination (and hence of transformation) of those elements. One might say that the computer enables a new kind of operational modeling of such structures, which is part of the reason for the rapid embrace of computer modeling.[18]

The world is more than just structured, however. It has a particular kind of structure, a tree structure—a very complex one, later made hyper-complex as the full implications of recursion and association began to be worked out. The tree structure was a surprisingly important concept. I once asked Herbert Simon what was the concept that most influenced his work, and he gave me two in answer: system, which I expected, and the tree structure, which I did not. The tree structure was a powerful mental model. It offered a new way to imagine *hierarchy*, enabling people to visualize, generalize, and formalize it. In addition, the model offered the hope of reconciling the static study of structure and the dynamic analysis of process. This reconciliation was difficult—was a tree structure a map of present relations or of developments over time?—but the idea of trees unfolding across space and time held great allure, connecting the conceptual scope of evolutionary theory to the practical power of computers and programs.

This point brings us to the next keyword: *process*. The social scientists who shared the bureaucratic worldview thought that the object of science was to describe systems in terms of the processes by which those systems maintained their internal equilibria, stayed in dynamic equilibrium with the environment, or acquired their present structures. Process, procedure, and sequence helped describe these adaptive operations. Hence the widespread fascination with descriptions of processes rather than states, with production systems, courses of action, strategies (sequences of moves), algorithms, heuristics, feedback paths, flow charts, and decision trees.

The processes that such scientists focused on at first were flows and transformations of energy, which in the interwar period were seen as the heart of the processes of adaptation and survival. During and after the war, flows and transformations of information came to equal, and even to exceed, those of energy as the crucial processes to study. This shift was linked to the enormous interest of the military during World War II and the Cold War in problems of communication, command, and control.[19]

Concern with communication processes leads us to the next keyword: *symbol*. High modern social scientists were fascinated by symbols. Many treated them as physical objects, presuming that manipulations of symbols entailed physical alterations in the world. To Herbert Simon, for example, the mind was simply a "physical symbol system," and he and his colleague Allen Newell considered this hypothesis one of their claims to fame.[20] Simon went so far as to claim that manipulating symbols was as "concrete as sawing pine boards in a carpentry shop."[21]

Some pursued this fascination with symbols through the study of symbolic logic, others through the study of codes, others through studies of the relationship between symbols, models, metaphors, and the things they represent, and still others through the attempt to develop new graphical forms to symbolize the processes, relationships, and time paths of systems: for example, flow charts, path diagrams, nomograms, circuit maps that represented logical operations, and even machines designed to simulate system functions, from Vannevar Bush's differential analyzer to the myriad devices developed during the depression to simulate the mechanisms of the economy. A vital innovation was Claude Shannon's 1938 discovery that physical circuitry could represent not merely counts of things but also logical operations and so could give physical form to abstract relationships among symbols.[22]

This interest in symbols and representations was a critical part of the development of systematic theory. To the high modernists, science was the product of the organization of facts into conceptual schemes, and the progress of science was due primarily to the development of more sophisticated, elegant, and parsimonious theoretical systems, not simply the discovery of new facts. In this view, systematic theory, especially as exemplified by sophisticated formal *models*, was the sine qua non of a true science. The great challenge was to create a rigorous empirical theory that would organize and so give meaning to brute facts while avoiding the enticements of the descent into the mists of metaphysics.

In keeping with this emphasis on theory, many high modernists (especially in economics, psychology, and linguistics) believed that the proper products of a reformed social science would be formal theoretical models. If a formal model was a science's ultimate destination, then its starting point was the empirical observation of a consistent relationship among a certain set of variables. A theory's purpose was to explain that relationship. A model, then, was a description of a concrete (physically realizable) system in terms of a theory. A theory, therefore, was not what one tested in an experiment, except indirectly: rather, one tested a model, which might be one of many ways of expressing that theory.

Some models were intended to be more general than such models of concrete systems. These abstract *structural models* were used to represent a general set of relationships common to different systems; they would thus have to be elaborated into models of specific, concrete systems to be put to an empirical test. An example would be the use of parallels between linguistic structures and kinship structures to draw higher-level, abstract conclusions about the necessary properties of such structures, as compared to the elaboration of those models to describe a specific group.[23]

Many researchers used the term *model* in a third way, thinking of models as *heuristics*, as opposed to as *representations of* or *formal abstractions from* concrete systems. Such heuristic models were much more like root metaphors or sets of basic assumptions than they were like concrete system models, and they were not couched in a formal language the way an abstract formal model would be.

One of the main reasons why modeling as a practice and communication as a subject were so interesting to high modernists were the promise they offered of *control*. Those who shared the bureaucratic worldview aimed quite specifically to understand the processes of control—the mechanisms that produce and maintain order—within systems, usually with the goal of enabling greater human power over them. High modernists thus shared the prevailing tendency among scientifically educated elites to be "rational reformers," in John Jordan's phrase, and were intent on bringing order and control through continuous management, but they firmly believed that improvements in practice required improvements in theory.[24] Hence, much high modern social science was abstract to the point of being otherworldly while remaining strongly instrumentalist in its orientation. Whether abstracted or applied, however, if nineteenth-century social energetics was concerned above all with *work*, this twentieth-century social energetics was deeply concerned with *power*, in the sense of authority, of command and control.[25]

That brings us to *synthesis*. The bureaucratic worldview exponents saw dichotomies (especially those that result in paradoxes) as signs of flawed thinking, and they desired to find a new perspective, a new conceptual scheme, a new set of tools and practices that would enable them to move beyond the longstanding divisions in science and society: mechanism and vitalism, mind and body, theory and practice, empiricism and rationalism, symbol and reality, choice and control, chance and design, freedom and order. Hence the debt to earlier metaphors, such as that of the human motor, that linked human and machine, and hence the postwar fascination with man-machine systems and various forms of cyborgism, both of which were central features of cybernetics and cognitive psychology. This desire for synthesis also contributed to their advocacy of the idea of the unity of science, a position most commonly associated with Vienna Circle positivists but that held great appeal far beyond that small group.[26]

Exemplars

System, structure, function, process, hierarchy, control, synthesis: these are abstract ideas. Abstraction was part of their appeal, for high modernists saw abstraction as a path to the universal. But such abstractions are difficult to understand today, just as they were challenging to apply then, without specific exemplars.

Three examples that illustrate the nature and significance of high modern social science and the bureaucratic worldview and that connect them to the institutions, values, and technologies of the organizational revolution are Chester Barnard and *The Functions of the Executive* (1938), Warren Weaver and his introduction to *The Mathematical Theory of Communication* (1949) and his essay "Science and Complexity" (1948), and J. C. R. Licklider and Robert Taylor on "Man-computer Symbiosis" (1960) and "The Computer as a Communication Device" (1968). Not only are these works important in their own right, but their authors also were powerful patrons who were able to advance their ideas with big dollars as well as powerful words. All three examples drew from, and contributed to the development of, powerful control technologies; thus, they reveal the two-way connections between these new technologies and new models of human thought and action.

Chester Barnard and the Functions of the Executive

Chester Barnard probably will be the least familiar name to most readers, though he was an extremely important figure in American social science and business. Barnard served as president of New Jersey Bell, head of the Rockefeller Foundation's General Education Board, president of the Rockefeller Foundation, and chairman of the board of the National Science Foundation. In addition, he was an active participant in the Harvard "Pareto Circle" of social scientists; friend and critic of several notable social scientists, including Elton Mayo, Talcott Parsons, and Herbert Simon; and author of *The Functions of the Executive*, a landmark analysis of the structures and functions of the modern organization.[27]

Barnard begins his study by noting that the "unrest of the present day" (the Great Depression) has attracted much scholarly attention but that this attention has not focused on the formal organization. "To me," he writes, "this failure of attention is like leaving a vital organ out of anatomy or its functions out of physiology." His goal is to describe the physiology of cooperative organizations, for without such analysis, we cannot understand the modern world, as formal organization is "a most important characteristic of social life, and . . . the principal structural aspect of society itself."[28]

Throughout his analysis, organizations are explicitly compared to organisms: "Systems of cooperation which we call organizations I regard as social creatures, 'alive,' just as I regard an individual human being, who himself on analysis is a complex of partial systems, as different from the sum of these constituent systems" (79–80). At other times, the analogy is to fields. "An organization is a field of personal 'forces,' just as an electromagnetic field is a field of electric or magnetic forces. The evidence of the effects, in both cases, is all that can be used to describe or define these forces." As a result, "an organization is a 'construct' analogous to 'field of gravity' or 'electromagnetic field' as used in physical science" (75). Seeing organizations as organisms and as fields are not divergent images to him because to say an organization is an organism is to say that it is a structured field of forces in equilibrium (6).

The qualities that matter for Barnard are organizational qualities. While "an individual human being is a discrete, separate, physical thing . . . for other and broader purposes, however, it seems clear that no thing, including a human body, has individual independent existence" (10). Hence, in his book, "persons *as participants in specific cooperative systems* are regarded in their purely functional aspects, as phases of cooperation. Their efforts are de-personalized, or, conversely, are socialized, so far as these efforts are cooperative" (16). This approach is justified because a group is "something more or different from the mere sum of the interactions between the individuals composing it. In this sense the group presents a *system* of social action which *as a whole* interacts with each individual within its scope" (41–42). As befits a president of a telephone company, Barnard sees communication as the core process of coordination, as that which makes an organization an organization and not a random collection of individuals.

Barnard views nature, society, the organism, and the organization all as having a hierarchic structure, with the different levels in the hierarchy being defined by scale and complexity (65). It would not be too much to say that he saw the world as having a structure similar to that of New Jersey Bell—and of the system it administered.

Following his discussion of scale, complexity, and organization, Barnard moves to an analysis of the "bases and kinds of specialization." He categorizes the different ways to specialize (finding five) and argues that "the effectiveness of cooperative systems depends almost entirely upon the invention or adoption of *innovations* of specialization." Thus, "in an important aspect, 'organization' and 'specialization' are synonyms" (135–36).

Barnard believes in free will as well as organizational control. Hence, perhaps the most important function of the executive is to reconcile these two divergent aspects of human life: "Cooperation and organization as they are observed and experienced are *concrete syntheses* of opposed facts, and of opposed thought and emotions of human beings. It is precisely the function of the executive to facilitate the synthesis in concrete action of contradictory forces, to reconcile conflicting forces, instincts, interests, conditions, positions, and ideals" (21). Linking his philosophy of organization, of science, and of democracy, he continues, "Free and unfree, controlling and controlled, choosing and being chosen, inducing and unable to resist inducement, the source of authority and unable to deny it, independent and dependent, nourishing their personalities, and yet depersonalized; forming purposes and being forced to change them, searching for limitations in order to make decisions, seeking the particular but concerned with the whole, finding leaders and denying their leadership, hoping to dominate the earth and being dominated by the unseen—this is the story of man in society told in these pages" (295–96).

Thus, Barnard presents the bureaucratic worldview already in full flower. The organization is an organism is a machinelike system of specialized processes united into a hierarchically structured whole by communications mechanisms. To manage such an organism is to effect a synthesis of seeming opposites, a task that only can be achieved by understanding the patterns of relationships that constitute the organization as a whole, for only then can the executive wield effective control over the seething system of exchanges (material and symbolic) of which he is both master and servant.

Warren Weaver, the Rockefeller Foundation, and the Sciences of Organized Complexity

Barnard closes with a quote from Plato's *Laws*, which says that though in human affairs chance is almost everything, "in a storm there must surely be a great advantage in having the aid of the pilot's art" (296). This comment foreshadowed a remarkable series of developments that led to making a science out of the "pilot's art," a science of decision, of coordination and control. One influential line of such work focused on the decision, the choice, and the strategy. Another closely related line sought to create radically new powerful ideas and tools for coordination and control through the scientific analysis of communication.

As head of the Rockefeller Foundation's program in natural science for over twenty years, president of the Sloan Foundation, head of the Applied Mathematics Panel of the National Resources Defense Council (NRDC) during World War II, and adviser to the Office of Naval Research (ONR), Air Force, and RAND after the war, Warren Weaver was unquestionably one of the most important patrons of science between 1930 and the early 1960s. His training was in field physics, but his interests and skills were far broader. His primary focus at the Rockefeller Foundation was in applying the tools, techniques, and concepts of modern physics to biology, which he believed was ripe for revolution in the 1930s. Guided by this belief, the Rockefeller Foundation became the great patron of experimental biology, biochemistry, and biophysics, paving the way for the creation of molecular biology and the DNA revolution. As head of the Applied Mathematics Panel (which he ran under the direction of Vannevar Bush, whom he knew well and whose differential analyzer he had funded before the war), Weaver sponsored work in a variety of fields, and many of the postwar leaders of operations research, game theory, and information theory, such as Phil Morse, Merrill Flood, and Norbert Wiener, did work for his panel.

Through his work on the Applied Mathematics Panel, Weaver became familiar with new work in electrical engineering on feedback, communications, and control, as well as with developments in electronic computing. Eventually, these connections led him to Claude Shannon, the gifted mathematician, electrical engineer, and unicycle rider of Bell Labs who had developed a new mathematical theory of communication. Weaver was taken with Shannon's theory, and he wrote a widely read introduction to Shannon's essays when they were published in book form in 1949.[29] At the same time he was working through the implications of Shannon's new information theory, Weaver wrote another well-known piece, titled "Science and Complexity," in which he argued that the day of the "sciences of organized complexity" was dawning.[30]

In these two essays, one can discern clearly Weaver's image of science and nature. In "Science and Complexity," he begins by sketching a broad picture of the development of science in which the physical sciences first prospered because they tackled "problems of simplicity" (536). Later, with the development of thermodynamics and the rest of modern physics, physical scientists discovered new statistical methods and so began to deal successfully with problems of "disorganized complexity" (537). The life and social sciences advanced as well, but not so dramatically. Their slower progress was because "the significant problems of living organisms are seldom those in which one can rigidly maintain constant all but two variables. Living things are more likely to present situations in which a half-dozen, or even several dozen quantities are all varying simultaneously, and in subtly interconnected ways" (536). The defining feature of such phenomena is that they all "show the essential feature of *organization*," and the relationship between their organization and their behavior is vital to understand (539).

To tackle the problems associated with "complexly organized wholes" required the development of new methods, tools, and concepts. Fortunately, "out of the wickedness of war have come two new developments that may well be of major importance in helping science to solve these complex twentieth-century problems." The first of these was "the wartime development of new types of electronic computing devices": "These devices are, in flexibility and capacity, more like a human brain than like the traditional mechanical computing device of the past" (541).

The second of the wartime advances was the " 'mixed-team' approach of operations analysis."

> These operations analysis groups were . . . mixed teams. Although mathematicians, physicists, and engineers were essential, the best of the groups also contained physiologists, biochemists, psychologists, and a variety of representatives of other fields of the biochemical and social sciences . . . Under the pressure of war, these mixed teams pooled their resources and focused all their different insights on the common problems. It was found, in spite of the modern tendencies toward intense scientific specialization, that members of such diverse groups could work together and could form a unit which was much greater than the mere sum of its parts. (541–42)

Weaver found such mixed teams so productive, and the powers of automatic computing so exciting, that he was tempted "to forecast that the great advances that science can and must achieve in the next fifty years will be largely contributed to by voluntary mixed teams, somewhat similar to the operations analysis groups of war days, their activities made effective by the use of large, flexible, and high-speed computing machines" (542). If that sounds like a good description of the actual course of development of postwar science, Weaver was part of the reason why.

To him, the thing that made these new tools and approaches so powerful was that they enabled the coordination of specialized labors into a larger whole: the sciences that studied organized complexity needed to be an organic (but differentiated) whole as well. The key to fashioning such wholes, among scientists or in nature, was communication. The triumph of modern science, Weaver writes, is in large part because "perhaps better than in any other intellectual enterprise of man, science has solved the problem of communicating ideas, and has demonstrated the world-wide cooperation and community of interest which then inevitably results." Indeed, in his view, science was "an almost overwhelming illustration of the effectiveness of a well-defined and accepted language, a common set of ideas, a common tradition. The way in which this universality has succeeded in cutting across barriers of time and space, across political and cultural boundaries, is highly significant" (543).

The creation of a science of communication was vital to this vision. In his introduction to *The Mathematical Theory of Communication*, Weaver stresses the importance of Shannon's new information theory: "The word communication will be used here in a very broad sense to include all of the procedures by which one mind may affect another. This, of course, involves not only written and oral speech, but also . . . in fact all human behavior."[31] Indeed, "the mathematical theory is exceedingly general in scope, fundamental in the problems it treats, and of classic simplicity and power in the results it reaches."[32]

What is this powerful theory? It is a description of the essential structures and functions of any and all communications systems, complete with new definitions of messages, signals, and information itself. Perhaps its most striking claim is that the fundamental processes of communication are ones of selection among alternatives: the information source selects a message out of a set of possible messages (or, more precisely, it selects a symbol out of a set of possible symbols, with messages being sequences of symbols).

Because the generation of a message is a process of selection among alternatives, "the word information in communication theory relates not so much to what you *do* say, as to what you *could* say. That is, information is a measure of one's freedom of choice when one selects a message." One consequence of defining information this way is that "the concept of information applies not to the individual messages (as the concept of meaning would), but rather to the situation as a whole." Weaver explains this connection several times: "Information is, we must steadily remember, a measure of one's freedom of choice in selecting a message. The greater this freedom of choice, and hence the greater the information, the greater is the uncertainty that the message actually selected is some particular one. Thus greater freedom of choice, greater uncertainty, greater information go hand in hand."[33]

This approach to information connects it to the physicist's concept of entropy: indeed, to Weaver the two are mirror images, with information simply being negative entropy. The conceptual and mathematical mirror-identity of information and entropy may be counterintuitive, but it is real, and, to Weaver, it means that Shannon's theory of communication helps get at the truly deep problems of organized complexity. To illustrate the importance of information as negative entropy, he quotes Sir Arthur Eddington: "Suppose that we were asked to arrange the following in two categories—distance, mass, electric force, entropy, beauty, melody. I think there are the strongest grounds for placing entropy alongside beauty and melody, and not with the first three. Entropy is only found when the parts are viewed in association, and it is by viewing or hearing the parts in association that beauty and melody are discerned. All three are features of arrangement."[34] Information is thus a property of the organization of a system, and communication is the essential immaterial something that connects the parts into an organized whole. The mathematical theory of communication, then, addressed all the problems of organized complexity— and brought the powerful tools of physical science to bear on them.

Weaver conceived of organic nature as a complexly organized system. He understood that system to have a tree structure—when asked to draw a picture of science, he drew a tree representing both the structure of nature and of science—with that tree structure existing in real time as well as conceptual space.[35] The fundamental process of communication was a stepwise, branching process of selection, of factoring messages two by two. He embraced the formal, mathematical analysis of organization and of organizational mechanisms, such as communication. He saw commonalities in operations research, game theory, information theory, cybernetics, systems analysis, decision theory, and in theories of the structures and functions of biological organisms, especially their processes of self-regulation and reproduction. He sought synthesis, and he saw such strong similarities between humans and machines that he was prepared to say that a machine could think—or, at a minimum, that it could model a thinking organism.[36]

Bush, Licklider, Taylor, and Man-Computer Symbiosis

The development of computer networks is not usually discussed in relation to the social sciences. Yet the high modern social science perspective on the human mind and body helped shape the basic assumptions and goals of the pioneers of interactive computing and computer networking.

The Internet has become a model as well as a medium. A large part of the unusual strength of the Internet's appeal as a source of models and metaphors comes from its seemingly organic qualities. Networks seem more flexible, more adaptable, less predictable than mere machines.

These organic qualities are not accidental. Rather, they were designed into the first interactive computers and networks by a group of people who held a specific set of ideas about how the human mind worked. Not only were ideas about minds reshaped by the encounter with computers, but also ideas about computers were reshaped by the encounter with ideas about the mind.

This new conception of humans emerged and spread from the 1940s to the 1960s. It was linked to a cybernetic vision of man-computer symbiosis and to the conviction that communication was fundamental to the problem-solving process. This vision inspired an influential group of patrons and researchers, almost all connected to the Department of Defense's Advanced Research Projects Agency (ARPA), to set a new agenda for computing, one that provided the machine with "organic" qualities even as a parallel program in psychology mechanized our understanding of our minds and bodies.

Four basic presuppositions characterized this view. The first two were that the human mind is an information-processing machine and that, although it is remarkable, the human mind-machine was being overwhelmed by the complexity of the problems and quantity of information of modern life. Together, these two ideas embody the model of man as a *finite problem-solver*. A third key idea was that complex problems could be solved through an intimate union of man and machine. Finally, the fourth idea was that communication processes were fundamental to the basic functions of this type of hybrid organism. All four of these presuppositions were new both to behavioral science and to engineering in the middle third of the twentieth century. All four likewise were clear expressions of the bureaucratic worldview.

According to this view, the finite nature of human reason explains why specialization is both so powerful and so frustrating. Only by specializing in the study of the world can we reduce the complexity of the problems we face. Doing so, however, often blinds us to alternative ways of viewing the problem at hand. Hence, to address complex problems we need to find ways to integrate the activities of many distinct, specialized problem solvers. In other words, the world is too complex for us to understand more than a few of the branches of the larger tree, yet it is essential to have a picture of the tree to make sense of the branches. Hence, we need to find some intellectual analog to the organizational technologies that modern industry uses so successfully to coordinate specialized physical work.

In Vannevar Bush's 1945 article, "As We May Think," one can see precisely this concern with the problem of integrating specialized knowledge to solve the complex problems of the modern world: "There is a growing mountain of research. But there is increased evidence that we are being bogged down today as specialization extends . . . Yet specialization becomes increasingly necessary for progress."[37] To Bush, and those (such as Douglas Engelbart, J. C. R. Licklider, and Theodore Nelson) who followed in his footsteps, the key to survival was "augmenting the human intellect," and the key to such augmentation was to find ways to unify man and machine. This hope was expressed vividly in J. C. R. Licklider's article "Man-computer Symbiosis," published in 1960. In this article, Licklider, a psychologist soon to be the head of ARPA's programs in both behavioral science and computer science, set out his goals: "The hope is that, in not too many years, human brains and computing machines will be coupled together very tightly, and that the resulting partnership will think as no human brain has ever thought."[38]

Licklider and his fellow high modern systems scientists believed such a symbiosis was possible because of the formal, functional similarities between minds and computers. In contrast to prewar work in the field of computing, high modernists consistently drew on the analogy between humans and machines to legitimate their symbiotic project. For example, in his landmark "First Draft of a Report on the EDVAC," John von Neumann refers to the different components in the computer system as "organs" and argues that the key elements of the computer system "correspond to the associative neurons in the human nervous system."[39] At an even grander level, W. Ross Ashby, Herbert

Simon, Norbert Wiener, and other high modernists constructed an abstract, functionalist theory of machines that treated systems of gears and levers, of neurons, of symbols, and of people in bureaucratic organizations all as fundamentally similar entities.[40]

This belief in the fundamental, functional likeness of mind and machine made it possible to think of a group of men and machines working together as forming a single meta-organism. The formal similarities between humans and computers made the construction of man-machine systems conceivable, while the real physical differences between them made it valuable.

This combination of functional likeness and physical difference formed the basis for an extensive program of empirical research into how to integrate man and machine. Some of the most prominent of such research efforts in the 1950s and 60s were those of John L. Kennedy and the staff of the Systems Research Laboratory at RAND, Doug Engelbart and his project on the "Augmentation of Human Intellect" at the Stanford Research Institute, and John McCarthy, Martin Greenberger, and the staff of Project MAC at the Massachusetts Institute of Technology.

In all these efforts, the challenge was to find ways to enable better communication between human and machine, for in this new view the problem-solving organism only extended as far as its communication lines. This view meshed perfectly with the widespread fascination with communication in postwar behavioral science, engineering, and natural science more generally. Indeed, one almost could say that the message was the medium in postwar science, as communication became a central problem in a range of fields, from cell biology and genetics to physics, engineering, and all the social sciences.

In the world of interactive computing, this fascination with communication first took the form of interest in enabling humans and computers to interact in real time via time-sharing techniques, graphic displays, and input devices. Once it began to seem possible to couple humans to computers in real time, Licklider and others in the ARPA community began to think of coupling humans via computers. Thus, seeing interactive computing as a step toward overcoming the limitations of the finite human problem solver was intimately connected to seeing the computer as a communication device.

Probably the most influential statement of this vision came from J. C. R. Licklider and Robert Taylor.[41] In their widely read 1968 article, "The Computer as a Communication Device," the connection between the bounded rationality of the human problem solver, the goal of cybernetic union, and the focus on communications is very clear. Following Herbert Simon, George Miller, and other cognitive psychologists, Licklider and Taylor state that "modeling, we believe, is basic and central to communication." Even more, communication is fundamental to thinking at all levels, whether it be communications among neurons, circuits, or humans.[42] Indeed, when minds communicate, new ideas inevitably emerge. The development of time-sharing, for example, was not just about "multi-access computing," it was about "machine-aided cognition" (to use the two different interpretations of what the MAC in MIT's Project MAC stood for).

The result was the reconception of the personal computer as a communications interface not only between people and machinelike "functionalities" but also among people. As is the way with powerful technologies, the result also was the reconception of more and more areas of life—and mind—via the model of the Internet and the tools of computer modeling.

Conclusion

In this chapter, I have taken two approaches to charting the rise of high modern social science and the bureaucratic worldview. First, the results of a survey of the flagship journals for anthropology, economics, political science, psychology, and sociology between 1925 and 1975 demonstrate that a new kind of social science emerged and flourished during this period and that at its peak it characterized a majority (or plurality) of work in all five fields. The chief finding of this journal survey is that there was a clear shift toward a novel approach to social science, one that employed new concepts and methods in the pursuit of new goals. In particular, there was a strong movement toward "systems thinking," modeling, and behavioral-functional analysis. This movement was very strongly correlated with an explicit embrace of theory, especially formal theory; it was strongly correlated with mathematization and quantification; and it was strongly (though not simply) correlated with the advent of new patrons for social science. It was also strongly *anti*-correlated with more traditional moral philosophy and with discussion of the content of current social or political issues, both of which found little support from patrons and which were decidedly nontheoretical and nonmathematical in conception and execution.

The concepts at the intellectual core of high modern social science may be captured by the keywords *system, structure, function, model(ing), behavior, process, communication, control,* and *synthesis*. These key concepts connect the work of leading scholar-patrons Chester Barnard, Warren Weaver, and the team of J. C. R. Licklider and Robert Taylor. Each of these figures was a strong exponent of this bureaucratic worldview in his research and writing; each was a "rational reformer" of the organization of science in order to better serve the needs of the large-scale bureaucratic organizations that supported the research he managed; each was intimately involved in the development or large-scale use of new communication and control technologies; each used such technologies as a heuristic model for understanding human behavior; and each used his position as a major patron to further work grounded in this outlook. Thus, these exemplars not only show the importance of these ideas and how they connect but also illustrate some of the connections between these ideas and their context: the continuing organizational revolution.

We now have a picture of the general shape of the branching tree of high modern social science. In the next chapter, we look at the ecosystem in which it took root: the new patronage system of postwar social science and the institutions that connected patrons, problems, people, and practices.

CHAPTER TWO

Patrons of the Revolution
Ideas, Ideals, and Institutions in Postwar Social Science

We should be problem-oriented, not discipline-oriented.

George Katona, 1954

One day in early autumn 1945, "a small group of faculty members in the social sciences . . . met for an informal discussion of the problems which would be raised for research in the social science fields in the University by the prospective passage of a bill to set up a National Science Foundation." The leader of this group, sociologist Talcott Parsons, wrote, "It was our unanimous opinion that the social sciences here at Harvard are not well equipped to deal with the situation which the passage of such a bill would bring about." What was the problem, in this group's view? "Almost any major field of research which is likely to sponsored by a federal agency is in the nature of the case bound to cut across the at present dominant divisions into departments and faculties."[1] In short, Harvard University would have to be reorganized for it to succeed in the new world of federal patronage for social science.

Parsons and his allies at Harvard were not the only ones who saw the emergence of new patrons for the social sciences after World War II as providing an opportunity for new ideas and institutions. Nor were they alone in their belief that these ideas and institutions necessarily would cross disciplinary lines. Rensis Likert and George Katona, for example, founded the University of Michigan's Institute for Social Research (ISR) in 1949 on the principle that "we should be problem-oriented, not discipline-oriented"—a belief the Office of Naval Research (ONR), one of their primary early patrons, actively encouraged.[2] Similarly, Herbert Simon and his colleagues at the new Graduate School of Industrial Administration (GSIA) at Carnegie Tech, also founded in 1949, believed that "the social sciences—weakened by a half-century of schisms among economists, political scientists, sociologists, anthropologists, and social psychologists—are undergoing at present a very rapid process of reintegration."[3] Simon and his colleagues agreed that while the "barriers to this kind of interdisciplinary work are great" within universities, extra-university patronage would enable them to turn the GSIA into an interdisciplinary social research institute.

Parsons, Likert, Katona, and Simon hoped that patrons would enable them to remake their fields and their universities. In may ways, their dreams came true: their institutional experiments at Harvard, Michigan, and Carnegie-Mellon were among nearly 250 interdisciplinary social science research institutes created in the first twenty years after the war.[4] Their intellectual programs swept their fields as well: anthropology and sociology were transformed by Parsonian structural-functionalism; political science underwent a behavioral revolution and psychology a cognitive revolution, both led by Simon, among others; and economists remade their discipline through two, roughly concurrent revolutions—the econometric and the macroeconomic (in which Simon also played a part).[5] In addition, all the social sciences were transformed by the statistical methods and survey research techniques developed by the staff of the Survey Research Center, one of the many branches of Michigan's ISR.[6] But Parsons, Simon, Likert, and their allies were more than a little surprised at the scale—and the consequences—of the changes they and their new patrons wrought. History teaches many lessons; one of the most important is to be careful what you ask for—you may get it.

Postwar changes in patronage for social science had wide ranging effects on both ideas and institutions.[7] To understand these effects, one must realize that the era saw two distinct, successive patronage systems for postwar social science, not one, as is commonly assumed. The first postwar patronage system played a major role in enabling the series of revolutions and interdisciplinary syntheses across the social sciences while the second encouraged the development of specialized concepts, techniques, and technologies *within* the disciplines. One unintended consequence of the rise of the second system was widespread concern among social scientists in the 1970s and 1980s that their fields were fragmenting.

Patronage and the Postwar Social Sciences

Discussions of postwar social science typically have been structured by disciplinary boundaries and oriented around intra-disciplinary debates. One reason for this disciplinary focus is that most histories of the postwar social sciences have been written by social scientists themselves, usually with the reform of their own individual fields in mind. This goal has led to some astute analyses of intellectual movements within disciplines, but it generally has not produced studies that look at similarities across disciplines, nor has it had much concern with patronage.[8] To take but one example, the state-of-the-art Cambridge history of *The Modern Social Sciences* structures its account of twentieth-century social science in America around the disciplines, and Dorothy Ross's essay is the only one in the entire volume that deals substantively with the effects of shifting patterns of support for the social sciences after World War II. "Funding" is listed in the index as being discussed on only *five* of the over 700 pages of the volume, and the only specific patrons mentioned are the Ford Foundation (2 pages), Rockefeller Foundation (7 pages), National Science Foundation (NSF) (2 pages), and National Institute of Mental Health (NIMH) (1 page).[9]

The historiography of the postwar social sciences also has been shaped by a focus on the rise or fall of the idea of social science as social critique. A very common theme thus has been the lamentation of the abandonment of political science's, or sociology's, or economics', or anthropology's critical project.[10] Such studies commonly link the rise or fall of a certain theoretical system or concept to the changing political climate or to specific political events, while a large, distinct literature discusses the role of social science in the political process.[11] Comparatively few studies, however, explore the changing nature and role of the patronage system. Such work is vital, however, for patrons and the institutions they support play crucial roles in mediating between the broader political context and the intellectual programs of individuals, as work on support for the *prewar* social sciences has demonstrated.[12]

There are some notable exceptions: James Capshew, Paul Edwards, Peter Galison, Roger Geiger, Craufurd Goodwin, Philip Mirowski, Dorothy Ross, Malcolm Rutherford, and Mark Solovey all have made valuable contributions in this line, with Mark Solovey's *Shaky Foundations* being a particularly welcome contribution.[13] Many of these scholars' studies focus on areas where the social sciences interacted closely with engineering (cybernetics, the study of man-machine systems, artificial intelligence, operations research), perhaps reflecting the influence of recent work on the postwar physical sciences and engineering, which has emphasized patronage and the institutional structures of Big Science almost to the exclusion of interest in the scientific concepts formed by individuals inhabiting those institutions.[14]

These scholars have made valuable contributions to our understanding of the postwar social sciences, though even their works have been characterized by a focus on changes *within* disciplines and a primary concern with the implications of patronage for the rise and fall of social science's critical project. These are important interests and concerns, but they are not the subject of this inquiry. Though this chapter critiques several of the findings and assumptions of the existing literature on postwar patronage for the social sciences, these critiques are offered as part of a remodeling project, not a "tear-down" and rebuild.

Ross, Solovey, and others have presumed that there was a single postwar patronage system that grew rapidly until the early 1980s; while on occasion new patrons entered the system and old ones left it, these changes did not alter the basic structure of the system. In their accounts, the big story is the arrival of the federal patron, with military patronage being distinct from other forms of federal patronage primarily in that it presented an extreme form of the general trend of the patron calling the piper's tune.

These scholars have emphasized four things regarding the postwar patronage system for the social sciences: (1) the relatively small scale of patronage for social science as compared with the physical sciences; (2) the consequent adoption of a "natural science" model for the social sciences, in hopes of winning greater support; (3) the avoidance by social scientists of politically controversial research projects, especially those dealing with sex, race, religion, and class; and (4) the pursuit of research projects that either supported the US government's foreign policy or could be sold as helping oppose communism and spread capitalist democracy.

All of these findings have merit, but they only tell part of the story. For example, it is unquestionable that the scale of support for the social sciences was much smaller than for the physical sciences, and a sizable amount of "physics envy" did ensue, especially among the program officers for the social sciences at the National Science Foundation.[15] However, postwar funding for social science research was much, much greater than it had been before the war, and this new funding was sufficient to alter the scale, scope, and nature of the research enterprise in many areas of the social sciences. Despite all the laments of program officers at the NSF, the social sciences, on average, *tripled* in size in the ten years between the late 1940s and the late 1950s, as measured by the number of members of the leading professional associations. Another telling sign of the new scale of extra-university funding for research is the sudden rise in the incidence of acknowledgments of research support in articles in leading journals: taking psychology as an example, barely one article of thirty in the *Psychological Review* offered such an acknowledgment as late as 1940, compared to twenty-one of thirty in 1950 and twenty-five of thirty in 1955.[16]

These changes reflected more than a new social convention regarding the acknowledgment of support. In many cases, articles citing extra-university research support reported the results of team research projects that could not have been undertaken without funding on a new scale. Such funding thus enabled big social science: research done by teams and oriented around the use of scarce, expensive resources, such as sophisticated instruments and large data archives. In short, the new scale of funding for the social sciences transformed them in many of the same ways that the physical sciences were transformed by their bigger dollars, and an emphasis on the larger sums won by the physical sciences tends to obscure this important fact.

An almost giddy sense of possibility was far more common among social scientists in the 1950s and 1960s than a jealous envy of the physicists. In field after field, the period from 1945 to 1970 was felt to be a "golden age" in which new capabilities and new concepts were matched by new prominence in public life. When Paul Samuelson wrote in the early 1960s that World War II had been every bit as much the economists' war as the physicists' (and perhaps even more so), it was both a playful poke at the pretensions of physicists and a proud proclamation of his profession's power.[17]

As regards the second point, it is true that the social sciences did embrace a "natural science" model in the postwar period, but this was nothing new. Despite some heartfelt protests against scientism in times of social crisis, especially the 1930s, the vast majority of leading social scientists in America have espoused a "natural science" model of what social science should be since at least the 1880s.[18] An aspiration to the status and rigor of natural science has been a constant among social scientists: what has changed is what they thought made natural science so powerful.

In the early twentieth century, most American social scientists believed that the keys to natural science's power were rigorous empiricism, the moral discipline to remain objective in the face of various forms of social pressure and personal weakness, and the mental discipline provided by the well-elaborated, integrated bodies of knowledge and practice we call disciplines. As Dorothy Ross has written, *the project* for the social sciences in America in the late nineteenth and early twentieth centuries was the creation of intellectually sound, institutionally stable disciplines.[19]

Social scientists who came of age intellectually between the 1930s and 1950s, however, tended to believe that what gave science its power was a combination of theoretical sophistication and precise experimental technique, with mathematics being the essential language of both theory construction and experimental analysis. Scientists, whatever the field, were people who built formal theoretical models and then tested them in controlled experimental situations, preferably with sophisticated equipment.[20] This shift in the understanding of the natural science model by social scientists, a shift toward the bureaucratic worldview and the high modern style, is what needs to be understood, not simply the near-universal embrace of some form of scientism.

Finally, the ties of social scientists to federal patronage and to private foundations leery of offending the public did influence some researchers to avoid inherently contentious topics. In this, they followed a long tradition: as Mary Furner showed in 1975, the original formation of the social science disciplines within American universities involved a battle between those who viewed social science as a platform for reformist agendas and those who viewed it as a value-neutral science.[21] The 1920s and 1930s likewise saw the Rockefeller Foundation plunge enthusiastically into new fields, such as public administration, only to back out again once the research they sponsored began to provoke alarm.[22]

Nevertheless, many of the researchers supported by the supposedly conservative, controversy-averse powers that be did extremely controversial studies. For example, both government- and foundation-funded social science research played a crucial role in supporting the Supreme Court's landmark *Brown v. Board* decision; launching the War on Poverty; altering debates about sexuality and gender roles; challenging traditional ideas about race, culture, and poverty; changing attitudes toward, as well as official policies regarding, birth control and population control; shaping economic policy, including crucial decisions regarding taxation and budget-making; and trying to spread democratic (well, anticommunist) ideas and institutions throughout the Third World.[23] Some of these research projects were more successful than others, of course, but all were extremely controversial, to say the least.

The extent to which the embrace of federal patronage led social scientists to censor their views is an important research question. It is not the only question, however; equally important is the extent to which such patronage gave social scientists the resources and public platforms necessary to speak to controversial social issues.

The key to untangling the implications of postwar patronage is to explore the fine structure of that patronage. Most accounts paint with a broad brush, assuming that one patronage system was built in the first decade after the war and that this system has remained largely unchanged in its basic structure and functions, despite its dramatic growth.

At the same time, authors of these accounts observe that the research institutions built during the 1950s and 1960s often collapsed or devolved during the 1970s and early 1980s. The Harvard Department of Social Relations was created in 1945, flourished into the 1960s, began to break apart in 1970 and finally dissolved back into departments of sociology, psychology, and anthropology in 1972. Carnegie Tech's Graduate School of Industrial Administration was created in 1949, flourished into the mid-1960s, when it was the model institution for the Ford Foundation's project to transform business education, and then began to turn into a very good, but not terribly unusual, business school. At first, Michigan's Institute for Social Research seems to be an exception, for it has grown so large as to boggle the mind. Even the ISR, however, still fits the pattern of growth, optimism, and a conscious pursuit of interdisciplinarity until about 1970, at which point concerns about stagnation, fragmentation, and the reestablishment of disciplinary boundaries became pressing issues.[24]

The standard accounts also include descriptions of the widespread belief among social scientists in the possibility of a unified, usable behavioral science in the 1950s and 1960s, contrasting this postwar optimism with the equally widespread lamentations of the 1970s and 1980s that the social sciences had become fragmented and politically impotent. Dorothy Ross, for example, describes the period from 1945 to 1970 as "Social Science Ascendancy" and 1970–2000 as "The Social Science Project Challenged" in her chapter in the Cambridge history of *The Modern Social Sciences*.[25]

Ross has a point: the big books in social science of the late 1940s to mid-1960s had bold, ambitious, declarative titles like *The Social System* or *The Political System* or *Foundations of Economic Analysis*.[26] The big books of the late 1960s to early 1980s, by contrast, had titles like *The Coming Crisis of Western Sociology* or *Maximum Feasible Misunderstanding*.[27] (My personal favorite is *Implementation: How Great Expectations in Washington are Dashed in Oakland, or Why It's Amazing That Federal Programs Work at All*.)[28] Such titles may not have represented the attitudes of all researchers, but they did reflect a broad change in the prevailing mood of social scientists.

Though Ross, Solovey, and others ascribe a vital role to the new patronage system of postwar social science in creating the period of ascendancy, they do not see the period of "disarray" as being the product of *changes* in that system. Rather, they appear to see it as a product of that system's continued growth, holding that the patronage system that fed the growth of the social sciences during the period of its ascendancy simply kept on growing until it created overspecialization. In the context of post-Vietnam doubts regarding science, liberalism, and American democracy, this continued specialization led to fragmentation and doubt rather than to a celebration of diversity.[29]

Support for the social sciences in the period of fragmentation and disarray, however, did not grow. Constant dollar funds for social science research began to decline in the very early 1970s, long before they crashed in the early 1980s. The age of interdisciplinary synthesis and "ascendancy," by contrast, saw not only astronomical growth and rapid specialization but also widespread optimism that a unified science of human behavior was being built.[30] Thus, one could argue that the social sciences fragmented when the pie began to shrink relative to the number of researchers (which continued to grow), not when the pie grew too large.

While this argument brings us closer to the truth, it only captures a part of it, for it still assumes that there was one system that either grew or shrank. The answer is a bit more complicated. The patronage system did not merely grow or shrink: it changed, and so did social science.

The Two Systems

Two distinct, successive patronage regimes shaped postwar social science. The first thrived from 1945 to the mid-1960s, while the second began to take shape in 1958, grew throughout the 1960s, and was clearly dominant by 1970. These systems overlapped between 1958 and 1970, a period of enormously rapid growth in funding for social science research (figures 2.1–2.3).

The First System

Program officers (and their advisers in academia) at several private foundations, the Social Science Research Council, and a range of military research agencies shaped the first patronage system. The main private foundations involved were the Carnegie Corporation, the Rockefeller Foundation, and the Ford Foundation (by far the largest of the three). The primary military research agencies sponsoring social science research were the Office of Naval Research,

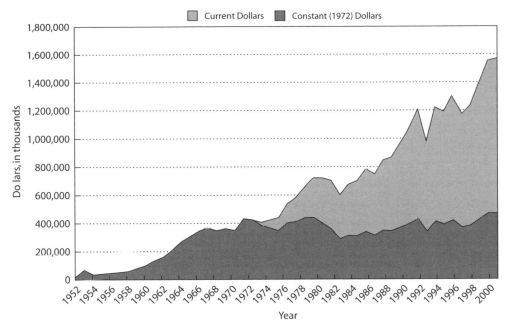

Figure 2.1. Federal funding for behavioral and social science research (BASS), 1952–2000. Data compiled from National Science Foundation, "Federal Funds for Research and Development, Detailed Historical Tables: Fiscal Years 1951–2001" (2001); National Science Foundation, "Federal Funds for Research and Development, Fiscal Years 1970–2001, Federal Obligations for Research by Agency and Detailed Field of Science and Engineering" (2001); Congressional Research Service, "Research Policies for the Social and Behavioral Sciences" (1986).

the Air Force Office of Scientific Research, RAND, and various units in the Army, including the Operations Research Office and several units that performed psychological research. Defense agencies often sponsored social science research as a part of large technical projects, such as the SAGE air defense system, making it difficult to determine exactly how much the military spent on social research.[31]

Despite their variety, these institutions were tightly interconnected at the top, with Merrill Flood, Rowan Gaither, Clyde Kluckhohn, Paul Lazarsfeld, Rensis Likert, Robert K. Merton, Robert Sears, Herbert Simon, and Ralph Tyler serving as program directors or members of governing boards and advisory councils for several agencies. These individuals all were at one time extremely successful clients of major patrons, whose advice many patrons and prospec-

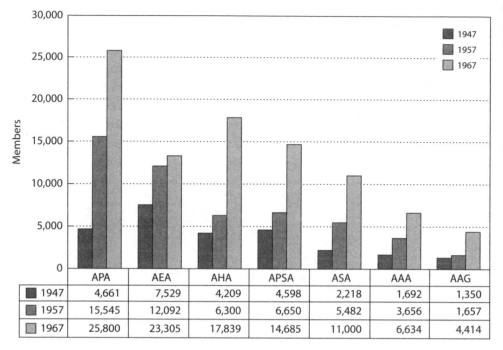

	APA	AEA	AHA	APSA	ASA	AAA	AAG
1947	4,661	7,529	4,209	4,598	2,218	1,692	1,350
1957	15,545	12,092	6,300	6,650	5,482	3,656	1,657
1967	25,800	23,305	17,839	14,685	11,000	6,634	4,414

Figure 2.2. Membership in the major social science professional associations, 1947–67. American Psychological Association (APA), American Economic Association (AEA), American Historical Association (AHA), American Political Science Association (APSA), American Sociological Association (ASA), American Anthropological Association (AAA), American Association of Geographers (AAG). Behavioral and Social Science Survey Committee of the National Academy of Sciences, *The Behavioral and Social Sciences: Outlook and Needs* (Englewood Cliffs, NJ: Prentice-Hall, 1969), 23.

tive clients soon came to seek. They played the essential role of brokers, linking patrons and clients in a well-articulated network of relationships.[32]

Take the career of Merrill Flood, for example. A mathematician fascinated with decision processes, Flood was a key figure in several military research agencies, and his ability to move easily from one agency to another reveals how widely the basic goals of the first system were shared. Flood ran the Fire Controls Research Office of the Army during World War II, sponsoring work by Norbert Weiner, Marston Morse, Albert Tucker, and other pioneers of operations research (OR) and cybernetics. After the war, he became the Chief Civilian Scientist of the War Department, in which capacity he sponsored research on OR, game theory, linear programming, and other mathematical approaches to

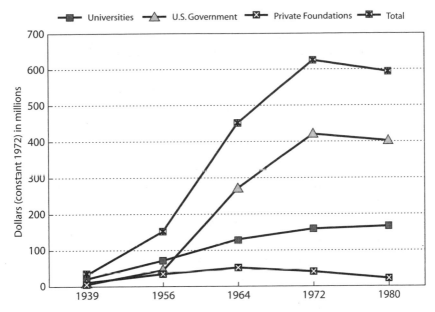

Figure 2.3. Behavioral and social science research (BASS) funding, by source, for selected years between 1939 and 1980. Data compiled from National Science Foundation, "Federal Funds for Research and Development, Detailed Historical Tables: Fiscal Years 1951–2001" (2001); Mark A. Abramson and Assembly of Behavioral and Social Sciences, Study Project on Social Research and Development, "The Funding of Social Knowledge Production and Application: A Survey of Federal Agencies" (Washington, DC: National Academy of Sciences, 1978); National Science Foundation, "Federal Funds for Research and Development, Fiscal Years 1970–2001, Federal Obligations for Research by Agency and Detailed Field of Science and Engineering" (2001); Congressional Research Service, "Research Policies for the Social and Behavioral Sciences" (1986); R. L. Geiger, "American Foundations and Academic Social Science, 1945–1960," *Minerva* 26, no. 3 (1988): 315–41; and Roberta Miller et al., "Research Support and Intellectual Advance in the Social Sciences," *SSRC Items* 37, no. 2–3 (1983): 33–49.

human behavior. Flood later moved to the Office of Naval Research, where he ran its Behavioral Models Project, an important source of funding for mathematical behavioral science, especially early cognitive psychology, in the mid-1950s.

Another example of these interconnections was "The Design of Experiments in Decision Processes," a conference held at RAND in the summer of 1952, which Flood helped organize. This conference was one of many where a mathematical, behavioral-functional, systems-theoretic approach to understanding decisionmaking was put forward as the centerpiece of an "interdisciplinary convergence." Many of the participants, such as Gerard Debreu, John Nash, John von Neumann, Frederick Mosteller, Howard Raiffa, and Herbert Simon either were or went on to be major figures in economics, sociology, or psychology. The conference also revealed the intertwining of sources of support for those working in this area: of the thirty-seven participants, twenty-one were sponsored by RAND, seven by the ONR, thirteen by the Ford Foundation, and three by the Cowles Commission for Research in Economics. In addition, seven speakers reported on research that had been funded by more than one of these four agencies. Both the ONR and RAND, for example, funded the University of Michigan's Behavioral Models Group, just as they supported Herbert Simon's group at Carnegie Tech.[33]

Brokers like Flood and Simon helped build a patronage system with the goal of promoting research that was mathematical, behavioral-functional, problem-centered, and interdisciplinary. The advocates of this type of social research usually called it "behavioralism" and began to refer to the social sciences as the "behavioral sciences." This behavioralist approach was an important part of the high modern vision for a new social science.

For the patrons—and clients—of the first system, there was no question that a reformed social science would be mathematical. Herbert Simon, for example, opened his book *Models of Man*, a collection of "Mathematical Essays on Rational Behavior in a Social Setting," by quoting Fourier's hymn to mathematics: "Mathematical analysis is as extensive as nature itself; it defines all perceptible relations, measures time, spaces, forces, temperatures . . . Its chief attribute is clearness; it has no marks to express confused notions. It brings together phenomena the most diverse, and discovers the hidden analogies which unite them . . . It seems to be a faculty of the human mind destined to supplement the shortness of life and the imperfection of the senses."[34]

The adoption of a behavioral-functional approach was a critical step on the road to mathematization. Behavioral-functional analysis did not lead inevitably to mathematical analysis (witness Talcott Parsons), but it did spur the development of mathematical social science in a couple of ways. First, the construction of a mathematical model requires simplification. The myriad details of individual identities and localities must be stripped away. The analysis of individuals as components of systems enabled such simplification, for it restricted the scientist's attention to a finite set of functions. Second, and similarly, a behavioral-functional approach allowed researchers to define theoretical terms in terms of systematic, measurable effects on a finite set of phenomena, making it possible not only to quantify relationships but also to describe behaviors as functions of each other (in the mathematical sense of the term).[35] Indeed, in the behavioral-functional view, an individual only could be known by his behaviors, and his behaviors could be known and identified only by their effects on the other elements of the system to which the individual belonged. An individual was precisely the sum of his functions.[36]

The final element in this high modern, behavioralist intellectual agenda was the development of formal theory and formal models. The high modernists sought to unite the empirical data gathering characteristic of American social science with the philosophically rigorous theorizing associated with European social thought. The objective was to create a rigorous empirical theory that would organize and so give meaning to brute facts while avoiding the enticements of descending into misty metaphysics. In the strongly empiricist context of American social science, this goal largely translated into a concerted effort to legitimize theory in the social sciences, usually done by embracing an active, manipulative experimentalism. To the high modernists, science was the product of the organization of facts into conceptual schemes, and the progress of science was due mainly to the development and experimental testing of more sophisticated, elegant, and parsimonious theoretical systems, not simply the discovery of new facts.[37]

These intellectual commitments were combined with support for problem-centered, interdisciplinary research. The high modern patronage system was shaped by the belief that work on meaningful problems necessitated a synthesis not only of different disciplinary perspectives but also of theory and practice. One consequence of this belief was the idea that research should focus on solving problems with tangible, testable, real-world implications in addition to having significance for theory. This "problem-centered" research was not simply applied research by another name; rather, the gradual elaboration of ever more sophisticated, ever more general, theory was crucial to both the patrons and the clients of this system. To men like Berelson or Parsons or Likert or Simon, theory only could advance by being put to the test in experimental situations, and practical problems only could be solved through the advance of theory. Thus, thinking of science as falling into one of two sharply defined categories, basic or applied, was a serious mistake.

The patrons of behavioral science believed that the proper way to conduct such interdisciplinary, problem-centered research was to create teams of researchers to work on projects funded by external contracts and grants at a small number of model research centers. In their minds, the ideal center would conduct both empirical and theoretical work and bring together specialists from several disciplines to work on projects funded primarily by the foundations and the military research agencies.[38] Concern with practical problems would be the spur to research, and research projects would move from the practical to the theoretical and fundamental before returning to the mundane once again. Happily, all three components of this vision reinforced each other: a focus on problems appealed to patrons with concrete problems to solve and, since real-world problems do not respect disciplinary lines, it simultaneously encouraged the "reintegration of the social sciences" that their leaders so ardently desired.

The high modernist agenda shared many features with that of the prewar Rockefeller Foundation (RF) and its affiliates. The parallels were not accidental: many of the leading patrons and brokers of the first postwar system had strong ties to the prewar RF and the Social Science Research Council (SSRC), which played an important role in promoting behavioralism in the 1950s and 1960s. There were, however, three significant differences between the prewar and first postwar systems: one of scale (and thus of influence), one of structure, and one of intellectual content and practice. First, the postwar system was much, much bigger in every way: dollars, people, institutions, and projects. Thus, while prewar efforts by the RF often were isolated and embattled, postwar ventures typically were parts of a broad, visibly successful movement. Second, the RF was by far the most important patron of social research before the war, with researchers (other than economists) having few alternatives other than it. There was no single, central patron after the war, however. Not even the Ford Foundation played the role of the prewar RF. And third, the emphasis on formal theory changed what the social sciences looked like in the postwar era.[39]

The Second System

The second postwar patronage system began to take form in 1958, its primary institutions being transformed in response to Sputnik. As these institutions grew in the 1960s, they came to have increasing influence on funding policy generally. Scientists and program officers at civilian, federal agencies, primarily the National Science Foundation and National Institutes of Health (NIH), shaped this second system. Significantly, both of these agencies had explicit missions to advance "pure," "basic," or "fundamental" science. By the 1970s, both were influenced strongly by officials who saw a sharp distinction between basic and applied science. For the program officers of the second system, applied social science meant the application or dissemination of existing social scientific knowledge, where the behavioralists had seen new, fundamental research as an essential part of solving practical problems. The NSF was more prone to this way of thinking, with the NIH being characterized by it to a lesser, but still significant, degree. Hence, though both agencies did fund applied social science—the NIH spent considerably more on "applied" than on "basic" social science—their understanding of "applied social science" was much narrower than the earlier definition of applied research as "mission-oriented basic research."[40]

The program officers at these agencies promoted research that would advance the several social sciences as disciplines, especially work that would lead to methodological or instrumental advance. As a group, they held no brief for or against any particular conceptual scheme, problem area, or philosophical stance, so long as the research being proposed was methodologically sophisticated.[41] The interest of these patrons in technical advancement was best expressed in their support for the development of computer modeling and simulation, the advancement of statistical technique, and the expansion and elaboration of survey research.

For example, mindful of the continuing need to prove social science to be truly scientific, both Harry Alpert and Henry Riecken, the first two program officers for social science at the NSF, chose to support research projects that promised mathematical or methodological advance.[42] While neither Alpert nor Riecken had a strong theoretical agenda or a decided preference for any particular discipline, the fields of social science that were most likely to use powerful mathematics and sophisticated instruments were cognitive psychology and various forms of statistical analysis. Hence, those fields won early support from the NSF; one of the first social science projects the NSF funded was "Mathematical Models for Behavior Data," a 1952 study by William Estes and C. J. Burke.[43] During the 1950s and early 1960s, if one encountered an article that explored a particular topic primarily as a means for testing a new statistical technique (as opposed to exploring it because of the intrinsic interest of the topic), odds were that the author's research had been funded by the NSF.[44]

Despite these differences, from 1958 through the late 1960s, the program officers at the NSF and NIH shared the same basic commitment to high modern social science that the leaders of the private foundations and military research agencies held. In fact, if citations of support are any guide, the NSF and

NIH were even more successful in their support of high modern social science during the late 1950s and 1960s than were the private foundations: according to the journal survey (see appendix), the NSF and NIH were markedly more likely to be cited in articles in the Core than in articles in the Outsider category (cited in 19% of Core articles vs. 4% of Outsider articles). The trend for the private foundations was similar but less marked: they were cited in nearly 9 percent of Core articles and just over 5 percent of Outsider articles. Similarly, the private foundations were just as likely to fund work in the Affiliates (or Margins) categories as in the Core or Inner Core, while the NSF and NIH were more likely to be cited in the Inner Core than in the Outer Core or Affiliates. In short, from 1960 to 1975, the NSF and NIH were even more closely associated with support for high modern social science than the private foundations had been at their peak period of the influence (1955–1965).

In addition, the NSF and NIH also put a greater emphasis on explicit behavioralism, sophisticated methodology, and modeling, which fit well with the rest of high modern social science during the 1960s, though modeling practice eventually diverged from its high modernist kin. Thus, the decline in high modern social science in the 1970s was influenced not only by the rise of the NSF and NIH relative to the private foundations but also by a shift within the NSF and NIH, as behavioralism, modeling, and methodological rigor became more highly valued even as they were less closely associated with a bureaucratic worldview.

Neither the military patrons of social science research nor the foundations disappeared after 1958. The effects of these first-wave patrons became increasingly localized within social science, however. The Ford Foundation, for example, abandoned its general program in the behavioral sciences at the end of 1957 and shifted its resources to the support of "area studies," reducing its impact on the mainstream of social science.[45] Likewise, the ONR continued to provide important support to certain strands of cognitive psychology in the late 1960s and 1970s, but its influence on psychology as a whole dwindled relative to that of the NSF and NIH, as can be seen in the shifting patterns of acknowledgment of support in leading psychological journals.[46] Military research in the social sciences became much more separated from mainstream research in the wake of Project Camelot, the Mansfield Amendment, and a general concern, shared by the left and the right (though for different reasons), that the military was funding research not closely related to its mission.[47]

This second patronage system overlapped with the first from roughly 1958 to the mid-1960s, but by the late 1960s the balance had shifted decisively in favor of federal, civilian support. This shift in the relative weight of the different patrons was accentuated by changes in the leadership of the relevant programs at the civilian, federal agencies. While many of the key program officers for the behavioral and social sciences in the NSF, NIH, and other federal agencies between 1958 and 1964 shared a philosophy similar to that of the high modernist leaders of the first system, a new generation of program officers took over between 1964 and 1972. They did not share—or were not able to implement—their predecessors' commitment to the high modernist cause. The primary exceptions were the various agencies associated with the War on Poverty, especially the Office of Economic Opportunity, which funded a wave of interdisciplinary research related to poverty from 1964 to the early 1970s.[48] Thus, in areas related to poverty studies, the period of overlap lasted until the early 1970s, at which point the second, discipline-oriented system began to assert itself.

One significant difference between the first and second systems was that although the NSF and NIH supported "applied" and "basic" social science, they generally saw the two as quite distinct categories, in contrast to the explicit support for mission-oriented basic research by the patrons of the first system. At the NIMH, for example, applications for applied and basic social research projects were sent to different study sections (peer review committees) for evaluation. In addition, at the NSF, support for applied social research often focused on technical advances, just as in its programs for basic research. A case in point is the NSF's Research Applied to National Needs (RANN) program of the 1970s, a response to congressional demands that basic research "pay off." The University of Michigan's Institute for Social Research won a three-year, $3 million grant under the RANN program: the proposal's chief innovations were to be advances in technique (statistical methods, data processing machinery and software) and advances in disseminating existing knowledge.[49]

The Reagan-era cuts in funding for social science research (especially notable in FY 1983) prompted a round of vigorous (and moderately effective) lobbying for federal support for social science research. The changing fortunes of the social sciences in the 1980s, combined with the multitude of laments regarding the "crisis" in every discipline (with the crises usually being associated with hyper-specialization—and with increased conflicts over how to divide a pie that never seemed to grow), spurred a revival of interest in interdisciplinary research. As the interdisciplinary ideal has been most strongly associated with mission-oriented basic research, it may be that the rhetoric of interdisciplinarity was more prevalent than the reality in the late 1980s and 1990s.

It is perilous to generalize about the intellectual implications of the second system, for its defining characteristic was openness to diverse theories, perspectives, and methodologies, so long as these varied approaches were validated by an accepted community of specialists. In addition, while the key figures of the first system promoted a high modernist, behavioral-functional approach quite deliberately, in both eras the patronage system influenced social research more through selective reinforcement than by mandate: to use an ecological metaphor, each supplied a different set of resources, but neither generated directly more than a few of the intellectual varieties that competed for those resources.

Nevertheless, one can discern some shifts in intellectual orientation that appear to be linked to these changes in patronage. For example, in the first period, researchers across a wide variety of fields studied systems as organized wholes, focusing on the mechanisms that integrated, coordinated, and stabilized such systems: in short, many shared the bureaucratic worldview. As a result, a great many social scientists were fascinated with communication, studying it as a mechanism of coordination and control. Classic examples include Parsons's *The Social System*, David Easton's *The Political System*, Karl Wolfgang Deutsch's *Nationalism and Social Communication*, and the incredible array of work in sociology, psychology, and anthropology that was influenced by cybernetics.[50]

In contrast, what leading social scientists in the 1940s to mid-1960s saw as wholes, later generations often saw as aggregations of individuals, as is evidenced by the rise of rational choice theories across a range of fields.[51] Herbert Simon, for example, shifted from studying how organizations affect the decisionmaking of their members to studying how individuals model the world in the course of solving problems.[52] Even the study of communication, the archetypical means of system integration, has come to be as much about individuals constructing representations—or about "memes" competing, in true Darwinian fashion—as about social coordination.[53]

Similarly, anthropologist Ben Orlove holds that anthropology has undergone a shift from "culture" to "knowledge" as the basic framework for study.[54] Studies using the culture framework, he argues, focused on understanding the beliefs and actions of individuals in terms of their place in a coherent system of meaning, whereas studies based on the knowledge framework treat individuals as things possessing sets of knowledge related to specific domains. These knowledge sets are understood to be isolable, dis-aggregable, and comparable across groups in ways that sociocultural functions never were, even in the most formulaic accounts.

Thus, it appears that the second system's demand for methodological rigor translated into a general tendency to be more accepting of "methodological individualism," in contrast to the systems-based framework of the high modernists.

Two Exemplary Patrons

One way to explore the structure and influence of these two patronage systems is to take a closer look at an "exemplary" patron for each. The most influential, and most representative, patrons for these two systems were the Ford Foundation and the NIMH, respectively.

The Ford Foundation

The Ford Foundation was easily the most influential single patron of the behavioral and social sciences in the 1950s. Between 1951, when it opened its doors, and 1957, when it decided to shut down its program in behavioral science, the Ford Foundation spent roughly $40 million on the "behavioral sciences," which is a larger sum than the NSF spent on *all the sciences put together* during that period. In addition, the Ford program in behavioral science was not the only way it supported social science; between 1963 and 1966, for example, it spent $35 million to reform business education, and the centerpiece of those reforms was to make business schools homes for social research.[55]

While it held a position of singular influence, it must be remembered that the Ford Foundation was but one of many patrons in this first postwar system. The Ford Foundation's resources and ambitions were greater than those of the other private patrons of the social sciences, but Bernard Berelson and the other program officers at Ford shared the belief that research should be oriented toward solving problems, that it should adopt mathematical and behavioral approaches, and that it should be interdisciplinary. Similarly, the foundation's internal pattern of coordination through interlocking committee memberships mirrored the structure of the patronage system for behavioral science as a whole.

The Ford Foundation was established in 1936, but its history really began in 1948, when the Ford family assigned a vast amount of their stock to the foundation in an attempt to lessen the bite of inheritance taxes.[56] This sudden influx of money meant that the foundation became the world's largest private philanthropy almost overnight, commanding greater resources than the Rockefeller and Carnegie philanthropies combined. The old organizational form and mission of the foundation were not sufficient to handle responsibilities of this magnitude, so the directors organized the Study Committee on Policy and Program to devise a structure and a purpose to direct its funds.

H. Rowan Gaither, who had just organized the new RAND Corporation in Santa Monica for the Air Force, chaired this committee.[57] Gaither's experience setting up RAND (with $1 million from the Ford Foundation) and his wartime experience as assistant director of the Radiation Laboratory at MIT had given him a thorough knowledge of Washington science policy circles. He knew that the Ford Foundation, despite its vast resources, could not compete with the federal government as a patron of research and development in the natural sciences. It would have to make its presence felt in other areas.

America's new global position, the emerging Soviet threat, and the stunning power of recent scientific advances (especially the atomic bomb) taught Gaither that the United States needed a better understanding of itself and the world beyond its shores. The foundation, he believed, could make its mark best by sponsoring programs intended to advance peace, democracy, and economic development. The Gaither Committee's report, therefore, advocated the organization of the foundation into five program areas, all quite broadly defined: "the establishment of peace, the strengthening of democracy, the strengthening of the economy, the improvement of education, and the better understanding of man."[58]

In 1951, political scientist Bernard Berelson was appointed program officer for the fifth of these areas. He soon set to work devising a strategy for the development of the *behavioral sciences*, a term coined in the late 1940s by a group of social scientists and biologists interested in social phenomena at the University of Chicago.[59] He chose "Behavioral Science" to describe his program area for several reasons, only one of which was the oft-cited desire to avoid the confusion of social science and socialism. He chose the term primarily because he desired to promote a particular kind of social science—mathematical, behavioral social science—and because he wished to avoid channeling its funds through the traditional disciplinary structures.[60]

After consulting with his advisers, Berelson circulated a draft plan in which he set forth his agenda. Agreeing with the Gaither Committee that "the critical problems which obstruct advancement in human welfare and progress toward democratic goals are today social rather than physical in character," Berelson stated that the "goal of the program is to provide scientific aids for use in the conduct of human affairs." Because the program's ultimate goal was the solution of social problems, not the mere increase of abstract knowledge, "the program [did] not fall within any one conventional field of knowledge, and traditional academic disciplines as such [were] not included or excluded. On the contrary, the program [was] interdisciplinary and inter-field "[61]

Berelson outlined a series of problem areas to focus research, such as "Political Behavior," "Values and Beliefs," "Social and Cultural Change," "Formal Organization," "Communication," and "Behavioral Aspects of the Economic System." All of these demanded interdisciplinary attack, Berelson argued, and they required the active leadership of foundation officers. In his view, the foundation should initiate action in these areas through specific programs and projects while it should respond to proposals in other areas. The foundation thus would not only select but also commission work in certain fields, conducting "inventories" of "tested propositions" in specific problem areas, such as organization theory, for example.[62]

The programs for each of these problem areas were to be developed through a network of interlocking advisory committees. Membership on one of these advisory committees was a sure sign of either existing or coming prominence. Membership on more than one such committee defined the inner circle of the social sciences: Paul Lazarsfeld, Robert K. Merton, Herbert Simon, Ralph Tyler, and Thomas Carroll—the organizational elite of the social sciences during the 1950s and 1960s—all served on at least four of these advisory committees, and their presence on multiple committees enabled them to give coherence to the foundation's overall program. Thus, the internal structure of the Ford Foundation's program in the behavioral sciences—a group of problem-oriented programs whose agendas were set by interlocking committees of advisers— mirrored the overall structure of the first postwar patronage system.[63]

The centerpiece of Berelson's proposed plan was the creation of an interdisciplinary "advanced education and training" institute to be "staffed by a small group of absolutely top men." This institute would have a permanent staff of ten to twelve, with a smaller number of visiting faculty, and it would train the leaders of the next generation of behavioral scientists. In Berelson's words, "the objective is to create *the* center for training and research in the behavioral sciences."[64] This "institute" eventually evolved into the Center for Advanced Study in the Behavioral Sciences (CASBS), another crucial node in the network of the postwar social and behavioral sciences. While some social scientists, such as Herbert Simon, never thought much of the CASBS, since it was not a research facility, it was, and still is, an important institution. The CASBS almost immediately became the place where the rising stars in social science came to write their big, synthetic works, and the consciously interdisciplinary nature of the institution contributed to such integrative projects.[65]

Beginning in 1953 the Ford Foundation also began to fund a series of six-to-eight-week intensive summer seminars for promising young faculty. The Social Sciences Research Council organized these seminars, and all promoted a mathematical, behavioral-functional approach to their topics.[66] Several of them explicitly focused on the interdisciplinary area of convergence identified by Herbert Simon as crucial to the development of a unified, usable behavioral science: decisionmaking in complex systems, today called the "systems sciences" by historians.[67] This was no accident, as Simon played a major role in organizing the summer seminar program and put together two of the most famous of the seminars, including one held at RAND in 1958 that became near legendary in the world of artificial intelligence and cognitive psychology.

These seminars, projects, and personal connections among patrons, brokers, and clients created a self-aware community. This community lacked a formal structure, but it made up for this lack with frequent, intense communication, collaboration, and competition; the use of a common mathematical language and behavioral-functional approach; and a shared sense of mission regarding the importance of understanding human actions as parts of complex social systems.

NIMH

The changing role of the National Institute of Mental Health offers an excellent example of the shift from the first to the second postwar patronage system for social science, for it was the most important single patron of the second system, if one disaggregates the various member institutes of the NIH. As the staff of the NIMH wrote in 1970, without exaggeration, "By 1965, NIMH was generally recognized by behavioral scientists in the country as the prime federal source for research and training support in psychiatry, psychology, sociology, anthropology, and behavioral biology."[68]

As the NIH was not an important patron of social science until 1958 (that year, again), historians have been prone to overlook its enormous role from that point forward or to treat its contributions as being similar to those of other, earlier patrons. Ellen Herman's excellent history of twentieth-century psychology is a case in point, for she scarcely mentions the NIMH as a patron of psychological research.[69] (She does discuss it in relation to public discourse about mental health and mental illness, which is important but not the same thing.) James Capshew's fine book *Psychologists on the March* is similar in this regard.[70] He notes that the NIMH became a significant patron of psychological research in the 1960s, but he treats its influence on the field primarily in terms of increasing the scale of funds available, and he does not study its agenda and impact in the same detail as he does other patrons, such as the NSF.

The NIMH, however, has been easily the single largest patron of psychological research in the world for at least thirty years, probably funding more psychological research in that period than all other patrons combined. In addition, the NIH and NIMH have had a distinctive influence on behavioral science. A subtle, but significant effect of NIH patronage for the behavioral sciences, for example, has been the reinforcement of a biomedical model for the methods and goals of social research, one that differed in some significant ways from the physical science or engineering model that is commonly assumed to be central to social science.[71] Broadly speaking, behavioral scientists in fields sponsored by the NIH have tended to think of themselves as physiologists of the body politic rather than as social engineers. The two concepts overlapped in some areas, but there were differences, especially as regards ideas about "therapeutic" interventions, with social experimentation often being understood as being rather like a clinical trial.[72] For example, Herbert Simon's applications to the NIMH emphasize the parallels between his methods of simulation, which model the cognitive behavior of individuals, and the fundamental processes of clinical diagnosis.[73]

This inattention to the role of the NIH, especially of the NIMH, in social science is partly due to the tendency to treat the federal government as a monolithic entity. The government is not a monolith, however, and all federal dollars are not the same. A good example of the importance of making distinctions between branches of the government is that in the 1960s, when cognitivists and behaviorists saw each other as opponents and rivals, the NIMH was a strong supporter of cognitive psychology while the educational research wing of its "grandparent" agency, the Department of Health, Education, and Welfare (HEW), was one of the few patrons of behaviorist research on conditioning.[74] When one distinguishes between NIH and HEW, then both the rise to dominance of cognitivism (thanks to the NIH's bigger dollars) and the persistence of selected behaviorist research programs (thanks to the continuation of independent funding streams from HEW) suddenly make sense.

Congress created the NIMH in 1946 to be the "National Psychiatric Institute," and its proponents intended it to be a center of support for training and research (in that order) in psychiatry and clinical psychology, as well as a sponsor of improved clinical psychiatric services. The NIMH's first director, Robert Felix, and all of the initial members of the institute's National Advisory Mental Health Council (NAMHC) were members of the Group to Advance Psychiatry (GAP), a group of psychiatrists who sought to promote psychodynamic approaches to psychiatric treatment.[75] Thus, the NIMH had a strong initial orientation toward improving clinical psychiatric services and an ambiguous relationship to biomedical research, both of which were unusual in the NIH.

Felix did have a strong interest in research, however, and he had an expansive vision of the kind of research that could improve clinical practice. A former colleague of Adolf Meyer at Johns Hopkins University, Felix had absorbed Meyer's belief that many mental health problems should be treated as public health problems.[76] As a result, Felix supported social research related to mental illness, especially studies of what might best be called the epidemiology of mental illness. He convened an external panel of social science consultants to advise him in this area, and its members—sociologists H. Warren Dunham and Robin Williams, anthropologist Margaret Mead, social psychologist Ronald Lippitt, and sociologist and social psychologist Lawrence Frank—steadily encouraged Felix to broaden his support of social science research. Significantly, all of these consultants shared a high modernist outlook, and all were interested in the culture-and-personality approach to understanding human behavior associated with Talcott Parsons, Clyde Kluckhohn, Henry Murray, and other members of Harvard's Department of Social Relations.[77]

Despite this encouragement, in the NIMH's first nine years its training and research grants were heavily oriented toward medical-psychiatric research: a 1958 review of the NIMH training program that categorizes the training grants by discipline of recipient does not even have a category for psychology, let alone the other social sciences.[78] Felix created an intramural Laboratory for Psychology in 1954 and provided crucial support for James Miller and Ralph Gerard's Mental Health Research Institute at the University of Michigan (the home of the new journal *Behavioral Science*), but up to 1958, both the NIMH's budget and the percentage of that budget devoted to social science remained small.

In 1958, however, Felix decided to expand the training program to include support for graduate training in psychology, including experimental psychology.[79] This tentative broadening of the NIMH's program met with great enthusiasm among academic psychologists, and it proved to be the proverbial camel's nose inside the tent. The NAMHC (which functioned as the NIMH's board of directors) gave this new program a ringing endorsement in 1959, even going so far as to encourage Felix to make the NIMH *the* leading source of support for "basic behavioral science."[80] Felix heeded this advice, and by 1964, 55 percent of NIMH principal investigators were psychologists while only 12 percent were psychiatrists.[81]

At the same time, the NIMH's budget began to grow with great speed, thanks to the post-Sputnik wave of support for scientific research. Over its first decade, the NIMH's budget had grown at a respectable rate, rising to $50 million in 1959. Between 1959 and 1964, however, that budget more than tripled, reaching $189 million. In part, the NIMH's rapid growth was simply a product of the growth of the NIH more generally, with Felix and NIMH riding on NIH director James Shannon's broad coattails. The NIMH's expansion also was due to its embrace of the research ethos of the NIH, however, an embrace manifested through its support of basic behavioral science.

The basic behavioral science Felix and the NIMH wished to support was exactly the kind of high modern social science Herbert Simon, Rensis Likert, George Katona, and other social and cognitive psychologists wanted to conduct: behavior*al*, not behavior*ist*, psychology. Felix and the members of the NAMHC were psychodynamic psychiatrists, as noted above, which meant that they believed in the mind, in consciousness, and in the ability of trained observers to gain insight into a person's mental processes by analyzing that person's verbal reports. Behaviorist strictures against all of the above seemed counterproductive to Felix and his staff, and the traditional behaviorist research program centering on animal psychology was far less appealing to them than was a focus on how humans think and learn.

Hence, Simon and other early cognitivists, such as George Miller, Bert Green, Carl Hovland, Jerome Bruner, James Miller, and Lee Gregg, found in the NIMH a generous patron with a strong interest in research on human thought processes. In addition, the NIMH actively supported the use of computers in psychology, creating a special funding program in the early 1960s specifically to expand access to computers for behavioral scientists.[82] The NIMH's generous patronage—which involved sums larger than the NSF and the private foundations put together—thus played a major role in promoting the cognitivist approach to experimental psychology, as well as in the development and widespread use of new research methods in the social sciences, such as computer modeling and simulation.

After Felix's departure in 1964, pressure grew to formalize the NIMH's grant programs and to align its work more closely with the biomedical ideal of basic research that characterized the NIH more generally. Though its program directors in the behavioral sciences continued to believe in interdisciplinary work, the NIH study sections (peer review committees) increasingly came to be defined by discipline—and even by specialty within disciplines. By 1980 there were *seventy* study sections, more than double the number of relevant disciplines found at most universities at the time.[83] The NIMH retained much more of a problem orientation than the NSF did, due to the interest of congressional sponsors in an eventual clinical payoff for the work it sponsored, but over time the distinction between research funded as "basic science" and that funded as "applied science" grew to be a formidable, though not impassable, barrier. As noted earlier, by the 1970s, different study sections administered by different program officers reviewed grant applications for basic and applied science.[84]

Scale and Structure

In addition to the shift in conscious emphasis from the high modernist embrace of mathematical, behavioral-functional analysis, mission-oriented basic research, and interdisciplinary research to an embrace of more discipline-specific innovation, the two systems also had indirect, and sometimes unintended, effects on the conduct of research and the structure of the social sciences. These indirect effects can be divided into two main categories: scale effects, which both systems produced in similar fashion, and structural effects, in which the two systems differed widely.

There are many ways a patronage system can affect a science. Patrons can provide differential support for or against various philosophical stances, methodological approaches, research topics, institutional or organizational forms, research practices, presentation styles, social values, concepts, theories, and research products, as well as for (or against) groups defined by social criteria such as race, gender, geographic location, or class. All of these effects may be the consequences of specific decisions patrons have made to support a certain project or class of projects, though many criteria are likely to be implicit rather than explicit factors in decisions, and some criteria will be much more important in one decision than in another.

A patronage system also can have effects that are largely independent of the choices of patrons regarding specific projects or classes of projects. The most important of these are the effects due to the general scale of support and to the structure of the patronage system. The most important scale effects in the postwar period have been specialization, which generally has increased as the size of the market (funding) for scientific research has grown, and expanded capabilities, especially via access to expensive equipment. The increasing focus of funding on technologies and techniques made it possible to employ large, expensive instruments, such as computers, and to create large, expensive data archives, as with large-scale social surveys. Such instruments and archives often became the foci for new disciplines, as in the case of computer science and survey research.

In addition to scale effects, a patronage system may have structural effects. Different patronage systems place responsibility for decisionmaking at different locations, rely on different kinds of intermediaries or brokers, and provide different avenues and media for interaction and exchange. The most important structural shifts in the postwar period had to do with the degree of centralization of decisionmaking regarding support and with the organization of the review process. For example, the second system saw a relative concentration of resources: there was roughly the same amount of money to be had in 1980 as in 1970, but the great bulk of it came from two sources rather than ten. Clients in the first system thus had more incentive to define their projects in ways that would appeal to multiple patrons and much greater freedom to ignore any single patron's demands than did clients during the second system. This is especially true as the NSF and NIH, in general, have tried not to fund work that lies within the other's purview, meaning that often there was only one serious option for funding for major projects after 1970.

In addition, the institutions at the center of the second system usually organized their grant programs and review panels along disciplinary lines, even though many individuals at the NSF and NIH continued to favor interdisciplinary work. Because of the structure of the review system, however, such interdisciplinary values tended to be muted, a fact recognized in repeated efforts to create "cross-cutting" funding programs at both the NSF and NIH in the 1990s and 2000s.[85]

The centralized, formalized structure of the review process in the second system reduced the significance of the broker—the extremely successful client of and adviser to multiple patrons—who had played an essential role in coordinating research in the first system. Certain aspects of the broker's role now have come to be played by university administrators (a dean or vice-president for research and his or her staff), whose job is to help their faculty win grants or contracts. These administrators, unlike the informal brokers of the early postwar period, typically are responsible for assisting faculty across a wide range of fields, in most of which they have no specific expertise. As a result, they are not responsible, explicitly or implicitly, for coordinating research in any specific field. In the present system, such research administrators are experts on the NSF or NIH grants process and advocates for their universities, not brokers seeking to link patrons and clients at multiple institutions into an articulated network.

Conclusion

In this story of patronage and the social sciences we see a familiar pattern, one common to a wide range of institutions and disciplines in postwar, especially Cold War, America. Time and again, interdisciplinary research centers were founded and flourished, often in close contact with similar centers, together forming a multidisciplinary, multi-institutional research community supported principally by military research agencies and by the Ford, Rockefeller, and Carnegie Foundations. Some prominent examples outside of the social sciences include molecular biology, solar system astronomy, materials science, and solid-state physics.[86]

Such centers and communities often were spectacularly productive, but they could be unstable as well. They were unstable because they depended on the shifting agendas of program officers at mission-oriented funding agencies and on the efforts of unusual individuals who could broker the interests of many groups. When agendas changed, or when such individuals moved on to other things, communities like the one oriented around Simon's interdisciplinary "convergence" or Parsons's systems theory usually dissolved into their component disciplines, perhaps changing them in the process, or they coalesced into new disciplines—such as computer science—that fit traditional university structures.

Centers like the GSIA, ISR, and Department of Social Relations repeated this pattern, either reproducing traditional disciplinary divisions within themselves, devolving into university departments, organizing into new institutes centered on powerful technologies or techniques, or separating from the university world entirely and reorganizing themselves as business enterprises. Which path they followed correlated well with the nature of their financial support: institutes with problem-oriented patrons tended to maintain an interdisciplinary focus; ones with discipline-oriented patrons of "pure science" tended to form discipline-based departments aligned with those patrons' interests; and institutes that centered on large, expensive instruments or large, expensive data archives tended to align themselves intellectually around those resources, creating new disciplines or new businesses in the process. Finally, institutes that supported themselves by producing marketable products or services tended to become businesses, as in the case of MITRE (a spin-off of MIT's Lincoln Laboratories), RAND's Systems Research Laboratory, and various polling organizations.[87]

Harvard's famous Russian Research Center (RRC) is a perfect case in point: in 1947 John Gardner of the Carnegie Corporation proposed the creation of "a program of research upon those aspects of the field of Russian Studies which lie particularly within the professional competence of social psychologists, sociologists, and cultural anthropologists"—a deliberate attempt to transform Russian studies into an interdisciplinary behavioral science. Its first director, Clyde Kluckhohn, also sought to "break away from the traditional disciplines (such as history, political science and literature) and to approach the subject with the help of the insights into human and social behavior gained by modern psychology, anthropology and sociology."[88] To borrow a phrase, it would be "social relations" in one country.

By 1971, however, the RRC's director Richard Pipes would write, "I must confess that personally I never have shared the belief that there exist 'keys' to the behavior of the Soviet Union and that by fairly ignoring a country's history and literature one could grasp the sources of its policies. I feel that my skepticism has been vindicated."[89] A reoriented, vigorously entrepreneurial RRC weathered the fiscal storms of the 1970s and early 1980s by finding a new network of patrons in business and government, selling itself on the basis of its unique archive of materials related to Russia, which remained an inherently interdisciplinary subject area. It succeeded so well that it outlasted its erstwhile subject of study, the Soviet Union.

Institutes like the RRC and the networks of extradepartmental, interdisciplinary centers in which they found homes have been among the most important sources of innovation in postwar science, physical, biological, and social.[90] They have thrived precisely because they provide an essential counterbalance to the insistent pressures toward specialization that pervade the academic world. The first postwar patronage system for the social sciences explicitly sought to create such centers, and it was successful in doing so. The second postwar patronage system placed a higher priority on technical and methodological innovation within disciplinary structures; as a result, such centers suffered unless they were able to redefine themselves around specific techniques or technologies.

This account is not intended either as a celebration of the first system or as a critique of specialization. A healthy academic ecosystem needs both strong disciplines and strong support for work that builds bridges between them. This account *is* intended to help us understand why the period of "social science ascendancy" and interdisciplinary synthesis felt so different from the post-1970, late modern period of crisis, challenge, and fears of fragmentation. In this case, following the money—and the ideals behind it—has provided some new answers and provoked some new questions that deserve scrutiny.[91]

CHAPTER THREE

The Magical Year 1956, Plus or Minus One

The holistic view of society as organism integrated from cell to nation depends upon the assumption that society, as an organization of living matter, is definable as a network of intercommunication.

A. F. C. Wallace, 1956

In 1956, George A. Miller published his famous article "The Magical Number Seven, Plus or Minus Two," one of several works from that year that heralded the coming of the cognitive revolution in psychology.[1] Also published in 1956 was A. F. C. Wallace's enormously influential article "Revitalization Movements," in which he applied a combination of culture-and-personality anthropology, cybernetics, information theory, Parsonian social systems theory, and physiological concepts (such as homeostasis and stress) to the study of major sociocultural movements, such as the birth and spread of new religions. And Walt Rostow's "The Take-Off into Sustained Growth," in which he introduced ideas that became central to modernization theory, including the notion that modernization was a universal process structured over time into distinct stages.[2]

These are but a small sample of the slew of ground-breaking works in the behavioral sciences published between 1955 and 1957, which also includes Simon's "A Behavioral Model of Rational Choice," "Rational Choice and the Structure of the Environment," and "The Logic Theory Machine"; Chomsky's *Syntactic Structures*, "Three Models of Language," and "Finite-State Languages"; Bruner, Goodnow, and Austin's landmark *A Study of Thinking*; Ashby's *Introduction to Cybernetics* (the first cybernetics textbook); Selye's *The Stress of Life* (which introduced the concept of stress to the general public—and to social scientists); Churchman's *An Introduction to OR* (one of the first operational research textbooks); Luce and Raiffa's *Games and Decisions* (the book that explained von Neumann and Morgenstern's game theory to social scientists); Parsons and Smelser's *Economy and Society*; Bauer and Inkeles's *How the Soviet System Works*; Dahl's *Preface to Democratic Theory*; Downs's "An Economic Theory of Political Action in a Democracy"; Lévi-Strauss's *Tristes Tropiques* (and, a wee bit later, in 1958, *Structural Anthropology*); the journal *Behavioral Science*, which launched in 1956; and, in reaction against all of the above, William Whyte's (frequently misinterpreted) *Organization Man*.[3]

Any three-year period will boast a good list of important works; what is exceptional about this period is how many foundational works, in how many fields, came out at that time. These were works that started long, productive lines of research, that received thousands of citations (when the average article receives less than ten), and that introduced terms—information, system, stress, organization, process—into the broader culture beyond academia. Moreover, it is remarkable how closely related the central concepts and assumptions of these foundational texts were, despite the wide range of fields involved. As should be no surprise, these were the concepts and assumptions of high modern social science, of the age of system.

These key works, published at the time of high modern social science's "takeoff," illuminate the interconnections among and implications of the core ideas at its heart. Despite their real differences (ones due to topic and discipline, among other things), these works shared significant assumptions. Among the most important of these was a set of linked beliefs that formed a new model of "man"—and it was always "man." This new model was not the perfectly rational, free to choose *homo economicus* of neoclassical dreams, nor was he the irrational, wholly malleable creature (*homo plasticus*, perhaps?) that both Skinnerian and Freudian psychology imagined. Rather, this was a model of man as *homo adaptivus*, the finite but inventive problem solver of bounded rationality. Homo adaptivus was a complex, hierarchically structured adaptive system, a system that existed within systems and that was itself a system of systems. Such systems, whether organisms or organizations, families or firms, nuclei or nations, were never to be seen as free-floating, isolated things. Every system, or subsystem, was bounded yet connected by ceaseless flows of energy and information. Information and energy crossed the boundaries between organisms and their environments and between systems and their subsystems (or co-systems), with the output of one being the input for another. Both the organism and the environment had characteristic structures and boundaries that were defined by these flows and conversions (or that channeled them, depending on one's point of view), to the point that an organism could be defined by its communication lines.[4]

High modernists described the basic mode of interaction with the environment of homo adaptivus in terms of choices or decisions that unfolded over time.[5] The organism, however, did not have sufficient knowledge of the world, or time to process the information it did have, to reach decisions that met an Olympian standard of rationality. Homo adaptivus lived in a world of information, yes, but that information was *expensive*. Hence, one of the most important sets of adaptive qualities was that which enabled an organism to do more with less information—to simplify a complex world. The researcher who studied homo adaptivus and his world necessarily had to do the same.

Of the many ways of simplifying the world, one of the most important was to channel an infinite array of possible responses into a small set of discrete choice options; another was to group and categorize the world according to a mental shorthand so that one thing (a symbol, a metaphor, a model) could represent many similar-but-not-the-same things; a third was to look for the first satisfactory result rather than the perfect one (to "satisfice" rather than to "optimize"); and a fourth was to develop a combination of special purpose mechanisms that would be triggered automatically by common situations and general purpose response mechanisms for dealing with everything else.[6]

The collected set of adaptive strategies of a community was its culture, for systems of knowledge and meaning are adaptive strategies, in this view, just as much as material technologies or social institutions. Adaptation was understood as success in maintaining internal equilibrium, with equilibrium usually seen as a dynamic thing achieved through homeostatic mechanisms.[7] Such mechanisms involved both interactions with the environment and internal adjustments to the stresses imposed by that environment.

Homo adaptivus usually was successful, or else it ceased to be, but this success was never perfection. For the fundamental feature of homo adaptivus was that he was a creature of *limited* capacities. He was a finite creature in an infinite world. Thus, homo adaptivus was a complex-but-limited adaptation machine, a bounded chooser, a finite problem solver. The limits to its powers were fundamental aspects of its existence. And, whereas traditional conservative views of human capabilities long have tended to emphasize our limits and fallibility, and traditional liberal views, our powers and perfectibility, this high modern view saw our very limits as the keys to our power and the path to our progress, if never quite our perfection.

Could humans ever escape their bounds? Unlike other creatures incapable of abstraction, we are able to represent, or model, this infinitely complex world symbolically. This process of symbolization enables simplification, analysis, comparison, compilation, forethought, design (including the design of tools as means to the ultimate end), and organized group action, especially group decisionmaking. Indeed, one could well say that language itself is a product of the limits of human cognition, as the inability to generate a word for every single individual object in the world forces us to create words for classes of things—and it is only through words for classes of things that we can communicate with people who have experience with other examples of that class but not with the specific object. By these means, cognitive and social, the human problem solver is able to move beyond local to more global maxima.

But even the most perfect model, explored by the most perfect human organization, necessarily will be imperfect. A perfect representation of reality would *be* reality and so would be *useless*. Our limits are the keys to our unique strengths, but displacing those limits does not make them disappear.

To understand the birth of this model man and its life across fields, it helps to begin with the works in psychology, cybernetics, and physiology that described homo adaptivus, the organism: Miller, Chomsky, and Simon on language, mind, and machines; Bruner, Goodnow, and Austin's *Study of Thinking*; and homeostasis, adaptation, and choice in Hans Selye's work on stress, Ashby's cybernetics, and Simon's behavioral models of choice.

Other social scientists scaled the model up from the organism to the organization, as in Churchman's analysis of the firm in *Introduction to Operations Research* or Downs's consideration of the party in "Economic Theory of Political Action in a Democracy." Indeed, Parsons and Smelser's *Economy and Society* and Wallace's "Revitalization Movements" broaden the scale all the way to society as a whole.

Another level of the system is the meta-level of the analysis of science itself, for a crucial consequence of seeing the cognitive simplification of the world as *the* distinctive human mental process was to see these simplifications—symbols, metaphors, models, conceptual schemes, paradigms—as fundamental to the highest expressions of human thought, including science itself. Science soon would be found to have a parallel structure, one that unfolded over time in a universal process that was close kin to those elaborated by Rostow and Wallace: a long period of preparation, followed by a sudden revolution, then by a takeoff into a new period of sustained growth based on a new paradigm.[8]

This process of revolution was driven less by the accumulation of new data than by the ability of cognitive structures—paradigms—to adapt to new stresses.

The Nature of Homo Adaptivus

The place to begin the study of the new model of man is the fertile nexus of cognitive psychology and cybernetic physiology. Both of these fields were themselves hybrids: the new cognitive psychology was a blend of information theory and Gestalt psychology, raised by operationalist experimentalist philosophy in a deliberately interdisciplinary home. The new cybernetic physiology was born of the physiological understanding of homeostasis (out of Henderson and Cannon) as joined with engineering concepts drawn from gunnery control and communications engineering, especially the concepts of feedback and control. Whitehead's organicism and Bridgman's operationalism were important resources for both psychology and physiology as well: although these two philosophical approaches were quite distinct (coming from different ancestral lines), elements of each could be harmonized. All of these fields received bounteous support from American military research agencies, which had developed a strong interest in communications, command, and control systems and in human-machine interactions.

The sum of this union of hybrids was the idea of a limited, error-controlled organic machine that always acted so as to preserve internal homeostasis (as best it could), with such adaptive actions involving the exchange, storage, or processing of information (and energy). Central to its adaptive mission was the ability to represent a complex external world internally and to do so in a way that was simple enough not to overtax the creature's limited capacities yet representative enough of the outside world as to be a valid guide to satisfactory, if never optimal, choices or actions. These root ideas can be seen clearly in the landmark works of 1955–1957 that began the "cognitive revolution": George A. Miller's "Magical Number Seven"; Noam Chomsky's "Three Models," "Finite-State Languages," and *Structural Linguistics*; and Bruner, Goodnow, and Austin's *A Study of Thinking*.

The Magic Number Seven

Information theory had spread like wildfire throughout experimental psychology in the early 1950s. One reason why information theory was accepted so eagerly was that behaviorist psychologists were used to seeing the body as a communication system. To a behaviorist, psychology was the study of correlations between stimuli and responses. Stimuli were readily equated with sources of messages, the brain with a switchboard, and nerves with transmitters and receivers.[9] Thus the step to information theory seemed a small one. Moreover, information theory promised to bring higher mental functions, especially language, within the ambit of experimental (rather than speculative) psychology.

George Miller, like other psychologists who worked at S. S. Stevens's Psychoacoustic Lab during World War II, the late 1940s, and early 1950s, was excited by this promise to both unify and mathematize psychology. Though he pursued Claude Shannon's information theory with the goal of reforming behaviorism, it soon took him down a different path. Information theory was a theory of messages, and to develop a complete theory of messages, Miller and others would find they needed a theory of the message generator—that is, a theory of mind. Homo adaptivus was an active chooser; mind (cognitive processes operating on internal symbolic representations) intervened between stimulus and response in the things that truly made us human.

In contrast to the behaviorist research program, the laboratory agenda set by information theory was intended to establish the specifications of the human communication system. The thresholds of sensation for various types of stimuli, the amount of information that could be perceived at one exposure, the channel capacity of the human nervous system, the amount of "noise" or interference present in various situations, the rate of transmission of signals and the length of communication routes in the human body—these were the important subjects of experimental study for those interested in information theory.[10] In short, it was less the amazing plasticity of human behaviors (which had fascinated Watson and Skinner) than the *limits* on that adaptability that were of central concern. This study of the limits of perception, representation, communication, cognition, and choice set a new experimental program for psychology (one that suited the needs of the military) and provided a new conceptual focus.

In his work at the Massachusetts Institute of Technology's Human Resources Research Laboratory (HRRL), Miller studied the ways in which the human operators of the SAGE computers related to their charges.[11] He examined how they perceived and processed the information the machines presented to them (in the form of arrays of lights or switches or symbols), and he attempted to discover the limits of the human organism's communications system. In "The Magical Number Seven," Miller synthesized the results of a series of studies of "absolute judgment." Absolute judgment refers to a person's ability to distinguish between one-dimensional sensory stimuli (e.g., their ability to identify a sound via differences in pitch or volume but not both). In the experiments he summarized, "the observer is considered to be a communication channel." The object of the experiments was to determine the "channel capacity" of the observer; that is, "the upper limit on the extent to which the observer can match his responses to the stimuli we give him."[12]

Miller reviewed the results of various experiments on absolute judgment and came to the conclusion that "there seems to some limitation built into us by learning or by the design of our nervous systems." This limit turned out to be surprisingly small: around three bits of information. A channel capacity of three bits of information means that we can identify roughly seven items, plus or minus two, if they differ only in one dimension (and are presented in isolation). The repeated discovery of limits on absolute judgment in the neighborhood of seven items caused Miller to remark, "I have been persecuted by an integer. For seven years this number has followed me around, has intruded upon my most private data, and has assaulted me from the pages of our most public journals . . . There is, to quote a famous senator, a design behind it."[13]

Miller does not stop with the discovery of the magic number seven. He notes the coincidence that the number of items that can be held in the immediate memory was also known to be around seven. At first, one might suspect that absolute judgment and the span of immediate memory might therefore be one and the same. Miller, however, quickly shows that this Pythagorean leap is mistaken. The span of immediate memory refers to one's ability to recall certain items from a much larger array of information. "Absolute judgment is limited by the amount of information [measured in bits]. Immediate memory is limited by the number of items. In order to capture this distinction in somewhat picturesque terms, I have fallen into the custom of distinguishing between bits of information and chunks of information."[14]

Miller reveals his discovery that "the span of immediate memory seems to be almost independent of the number of bits per chunk."[15] (In another work, he described the immediate memory as being like a purse that can only hold seven coins, but that does not care whether they are seven pennies or seven silver dollars.)[16] This curious feature of immediate memory is crucial, for it allows us to "increase the number of bits of information that [our memory] contains simply by building larger and larger chunks, each chunk containing more information than before."[17] To give this successive "chunking" of information a name, Miller borrowed from communications theory the term *recoding*.

Though the term *recoding* was taken from information theory, Miller's use of the term was strongly influenced by his early experience with computers. For example, one of the key experiments on which he based his idea of recoding was the "chunking" of sequences of binary digits into larger units, which was exactly what the computers (and their operators) he was working with did every day. Indeed, Miller himself has stated that the idea of recoding came from watching human operators in the control rooms observing complex arrays of lights and switches and "recoding" these arrays in their minds into easily recognizable patterns of light and switch arrangements.[18]

A Study of Thinking

Recoding was an internal cognitive process, an adaptation designed to aid finite beings in their attempts to deal with an infinite world. It was also an act of creation, for every abstraction is an invention, as Bruner, Goodnow, and Austin argued in their monumental book, *A Study of Thinking*.

The authors begin with an observation: "The past few years have witnessed a notable increase in interest in an investigation of the cognitive processes— the means by which organisms achieve, retain, and transform information."[19] Where did this revival come from? In their view, it partly came from the "recognition of the complex processes that mediate between the classical 'stimuli' and 'responses'" (vii). This recognition led to a "mediation model" instead. "As Edward Tolman so felicitously put it some years ago, in place of a telephone switchboard connecting stimuli and responses it might be more profitable to think of a map room where stimuli were sorted out and arranged before every response occurred, and one might do well to have a closer look at these intervening 'cognitive maps'" (vii–viii). Such a cognitive map is, of course, a mental model of the world. Thus, the "mediation model" of mind was a model of the mind as model maker.

Another input into the "revival" was information theory. While information theory initially seemed to fit traditional "S-R psychology," they note, "the inputs and outputs of a communication system, it soon became apparent, could not be dealt with exclusively in terms of the nature of these inputs and outputs alone nor even in terms of such internal characteristics as channel capacity and noise" (viii). Just as Miller had found in "The Magical Number Seven," "the coding and recoding of inputs—how incoming signals are sorted and organized—turns out to be the important secret of the black box that lies athwart the communication channel" (viii).

To these two sources (the "mediation model" and the recoding of informa-tion), Bruner, Goodnow, and Austin add a third: the rise of a new personality theory, derived from Freud (and improved on since his day, they add a bit hastily). Citing Anna Freud's *The Ego* and Gordon Allport's *Personality* (both published in the 1930s), the authors argue that the work that came to be called the "New Look" in perception soon evolved into a search for links between "general laws of perception and cognition on one side and general laws of personality functioning on the other" (viii). These three sources of change had been brought together at the Laboratory of Social Relations at Harvard in its five-year "Cognition Project," funded by the Rockefeller and Ford Foundations. *A Study of Thinking* was the result.

Yet this constellation of inputs, intellectual and institutional, needs one further addition: the assumption that the human mind, while a complex and powerful thing, is a finite machine tasked with making sense of an infinite world. Again, it was the limits of the mind that made a mediation model nec-essary and that made recoding and symbolic representation essential.

But how were such mental maps created? Miller had looked at the recoding of information into symbolic packages; Bruner, Goodnow, and Austin focused on a related, but distinct, step in the cognitive process: categorization, the sorting of things into groups or classes, which they saw as the fundamental process involved in "conceptualization." "This book is an effort to deal with one of the simplest and most ubiquitous phenomena of cognition: categoriz-ing or conceptualizing . . . We have sought to describe and in a small measure to explain what happens when an intelligent human being seeks to sort the environment into significant classes of events so that he may end by treating discriminably different things as equivalents" (viii). Categorizing serves many purposes, but the first they list should by now be familiar: "By categorizing as equivalent discriminably different events, the organism *reduces the complexity of its environment*" (12).

Without the basic assumption that the human mind is limited, this process is unintelligible. We have, they note, the sensory capacities necessary to dis-tinguish an infinite array of phenomena, to perceive every event, every object, as a singular thing. But, like Jorge Luis Borges's "Funes el Memorioso," who could remember every leaf on every tree in every season of every year, we would be "slaves to the particular" if we could not categorize (1). "To catego-rize is to render discriminably different things equivalent, to group the objects and events and people around us into classes, and to respond to them in terms of their class membership rather than their uniqueness . . . We map and give meaning to our world by relating classes of events rather than by relating individual events" (1, 13).

This presupposition of limits leads to a model of humans—and minds—as creators: "The process of categorizing involves, if you will, an act of invention" (2). The highest form of this conceptual process, to the authors, is science, and "the development of formal categories is, of course, tantamount to science-making" (6). They continue in language that Thomas Kuhn later would develop and make familiar to every historian of science: "Science and common-sense inquiry alike do not discover the ways in which events are grouped in the world; they invent ways of grouping. The test of the invention is the predictive benefits that result from the use of invented categories. The revolution of modern physics is as much as anything a revolution against naturalistic realism in the name of a new nominalism" (7). Science is thus not only a uniquely human activity; it also is a *characteristically* human activity, one that emerges from the necessary structures, functions, and limits of our minds.

Like other high modern works of the cognitive revolution, *A Study of Thinking* is deeply concerned with cognitive processes. The authors are concerned with sequences of mental operations that transform singular objects into exemplars of classes of things. These sequences are structured and so may be termed *processes* (rather than chains of historical accidents), and it is the structure of these processes that Bruner, Goodnow, and Austin seek to discover. Their goal is not to explain concepts but "concept attainment," in particular, to elucidate the "search strategies" that enable one to create instrumentally useful categories. Here one should note that *strategy* is a weighty word, for, as the authors explain, the importance of longer sequences of actions was suggested to them by J. Robert Oppenheimer, referencing the work of his Institute for Advanced Study colleagues John von Neumann and Oskar Morgenstern on strategies in game theory. The concept of the strategy as an ordered sequence of operations, taken as a unit, soon became vital not only to game theory and operations research (OR) but also to computer science, as from the seed of "strategy" the broad tree of the "program" swiftly grew.[20]

On another level, thinking as *processing* involves not only the transformation of mental objects but also the transformation of the systems of relationships that make that mental object intelligible. The "information processing model of mind" that was and is central to cognitive psychology thus was notable for its focus not just on information but also on process, model, and mind.

To Bruner, Goodnow, and Austin (and to almost all cognitive psychologists) the most uniquely human of the cognitive products of the ubiquitous processes of categorization, representation, and transformation was the creation of shared symbol systems—languages—for to them, every word was a model and a metaphor, a simplified representation of an infinite world, a bit of sound or text that grouped an array of distinguishable individualities under a common rubric that highlighted their similarities while pushing their differences into the empty spaces between words.[21] Every sentence was not just a chain of such symbols but a structured sequence, a verbal strategy, a cognitive course of action, a linguistic program. But how were these words connected into an intelligible structure?

The Grammar of the Mind

In a series of works published between 1955 and 1957, Noam Chomsky outlined how a "finite-state machine" (the human mind, or, as he later would argue, a Turing machine) could connect words into structured sequences. Chomsky argued that the statistical analysis of communication was insufficient to account for the complexity and regularity of natural language.[22] He held that when people hear or read something, they remember only a "kernel" sentence. A kernel sentence is the simplest declarative form in which the idea of the sentence can be expressed. This kernel sentence is then transformed according to various rules to suit the situation at hand. Chomsky drew from this analysis the conclusion that grammar, not wording, was the key to meaning. Syntax governed semantics. In addition, he concluded that syntax was rule-governed and analyzable through symbolic logic.

Chomsky then tried to discover what the rules for sentence generation must be. In order to do so, he defined a language as a certain set of possible sentences. The set of rules that specify that particular set of sentences—all and only the syntactically correct sentences in a language—Chomsky called a grammar. Each language's grammar was derived from a universal "transformational grammar" (the existence of which is proven, in his view, by our ability to translate between languages).[23] The key was to realize that in sentences, words are not simply chained together in a linear, probabilistic sequence; such sequences, Chomsky famously said to George Miller, "will never converge on English."[24] They demanded too much of homo adaptivus's limited processing power. Rather, grammars formed fundamental structures that related words to each other, especially recursive, hierarchical structures that enabled us to "nest" meanings as well as to chain them, and knowledge or possession of a grammar enabled our limited, all-too-finite minds to work extraordinary feats of symbolic representation and transformation.

Words—categories, abstractions, models, metaphors, and symbols—were not simply generated by the mind, they were interrelated, connected, processed and transformed by being brought into new relationships with other words. And these transformations enabled us to communicate and to act in a world so infinitely diverse, so fantastically singular, that it otherwise would beggar our meager mental capacities.

As was so common among high modernists, Chomsky's linguistic homo adaptivus blurred the lines between human and machine. Chomsky's definition of a grammar made it almost identical to a computer program, as both involved the manipulation of symbols according to the sequential application of a set of logical rules. Recoding clearly was a similar process. The chunks of information produced by recoding were symbolic representations formed, organized, and related to one another according to a set of rules akin to a grammar—or a program. Moreover, if each language's grammar was like a program, then the transformational grammar must be analogous to the operating system of the human mind. This link between programs, grammar, and mind was made explicit by Chomsky's later proof that a transformational grammar is formally equivalent to a Turing machine and thus, to many cognitive psychologists, to a mind.[25]

Bounded Rationality and the Limited Chooser

The particular program that Miller, Bruner, Chomsky, and other experimental psychologists had in mind when they talked of mind was Herbert Simon and Allen Newell's Logic Theorist (LT).[26] In a telling convergence, at the same conference that Simon and Newell presented their first paper on LT, Chomsky delivered a paper outlining his new approach to language generation and Miller presented "The Magical Number Seven." To Simon and the rest of the audience, proto-cognitivists and proto-artificial-intelligence researchers all, it was clear that a revolution was in the making.

The path that led Simon to LT indicates the breadth of this convergence on a new model of man, for he came to psychology, communication, and computing from quite a different direction.[27] Simon had been trained as a political scientist and economist, not as a psychologist; his work had focused on decisionmaking within organizations rather than cognition within individual minds. But to Simon and his fellow high modernists, not only was an organism an organization, an organization was also an organism. And all were adaptation machines. Simon's theories of organizational behavior thus were intimately connected to the emerging model of humans as a creatures of bounded rationality: indeed, he coined the term *bounded rationality* in his 1958 collection of essays, *Models of Man*, though the principle was evident in his work well before he used the term.[28]

Simon's work on administrative decisionmaking had revealed to him the importance of the principle of bounded rationality and the centrality of the problem of choice, and it had taught him to see both the individual and the organization as decisionmaking machines. From cybernetics and servo theory he now added the ideas that organisms, organizations, and adaptive machines were functionally equivalent, not merely similar; that feedback was an essential component of all adaptive systems, organic and mechanical; and that adaptive systems could evolve enormously complex behaviors by "nesting," rather than chaining, simple behavior mechanisms. The term *evolve* is important, for, in Simon's view, the mechanism that produced both adaptive behaviors and their organization into a hierarchical system of behavior was a process akin to natural selection. Finally, from Gestalt theory Simon learned to think of learning and problem solving as processes of cognitive adaptation: creatures adapt to their environment by learning to construct simplified mental models of that environment, models that serve not only as the reference points for decisions as to how to achieve the organism's goals but also as the basis for defining the goals themselves.

All of these ideas were grounded in two essential beliefs: the principle of bounded rationality and the idea that the world is a system. The overall product, the model of homo adaptivus, reflected those ideas throughout: the human in this model is a simple, error-controlled creature that, in the course of its otherwise random encounters with an environment so complex as to be incomprehensible in its entirety, learns to construct simplified models of its environment, models that enable it to satisfy its goals, though not to achieve them in optimal form.

The first major milestones along this path were a pair of articles in which Simon outlined a new approach to rational choice, "A Behavioral Model of Rational Choice," and "Rational Choice and the Structure of the Environment."[29] In his collection *Models of Man*, issued shortly after these articles were published (an indication that he believed he had reached the end of one period in his career and the beginning of another), Simon introduced these two articles by noting that "the publication in 1945 of von Neumann and Morgenstern's *Theory of Games and Economic Behavior* has attracted enormous attention to the theory of rational choice. This has been reinforced and amplified by parallel developments in mathematical statistics . . . which have reinterpreted the theory of statistical tests as a theory of rational decision." While these developments in the sciences of choice were of "the greatest importance," Simon argued, "the approach taken in the theory of games and in statistical decision theory to the problem of rational choice is fundamentally wrongheaded." It was wrong in "precisely the same way that classical economic theory is wrong—in assuming that rational choice is choice among objectively given alternatives with objectively given consequences that reflect accurately all the complexities of the real world." It was wrong, in short, in "ignoring the principle of bounded rationality, in seeking to erect a theory of human choice on the unrealistic assumptions of virtual omniscience and unlimited computing power."[30]

It was time, therefore, for a "fundamental change in our approach." It was time to "take account—and not merely as a residual category—of the empirical limits on human rationality, of its finiteness in comparison with the complexities of the world with which it must cope." The key to making this fundamental change was to study the psychology of rational behavior; that is, to take into account the psychological properties and limitations of the individual chooser. "The alternative approach employed in these papers is based on what I shall call the *principle of bounded rationality*: The capacity of the human mind for formulating and solving complex problems is very small compared with the size of the problems whose solution is required for objectively rational behavior in the real world—or even for a reasonable approximation to such objective rationality."[31] In keeping with this outlook, he wrote in "A Behavioral Model," the task was to "replace the global rationality of economic man with a kind of rational behavior that is compatible with the access to information and the computational capacities that are actually possessed by organisms, including man, in the kinds of environments in which such organisms exist."[32]

Simon begins "A Behavioral Model" with a discussion of rational choice that draws from game theory, moving from this "unrealistic" model of human behavior to one resting on less "heroic" assumptions. The result is a model of great simplicity and generality. He describes the "choosing organism," for example, in terms that could apply to any organism in any situation, analyzing it in terms of the set of behavior alternatives available to it, the subset of those behaviors it actually considers, the possible future state of affairs, and the payoffs for the organism attendant on these various possible future states.

He then sketches the basic game theory and probability rules that have been developed to guide the organism's choices. He does so only to point out their limitations, however, noting that these "classical" concepts of rationality make "severe demands" on the organism. The organism, for example, must be able to "attach definite payoffs (or at least a definite range of payoffs) to each possible outcome." There is thus no room in these models for "unanticipated consequences" or for incomplete orderings of preferences. To Simon, such narrowness invalidates these theories: "There is a complete lack of evidence that, in actual human choice situations of any complexity, these computations can be, or are in fact, performed."[33] As he wrote to the psychologist Ward Edwards, "About the kindest thing I can say about the maximization model is that: humans being the compliant creatures they are, you can sometimes induce them (by defining the game that way in real life or the laboratory) to behave more or less like the model of rational man (if you make the choice-situation sufficiently simple-minded)."[34]

In his view, the choosing organism is limited not only by the external constraints on its choices that the game theorists recognized but also by internal constraints: "Some of the constraints that must be taken as givens in an optimization problem may be physiological and psychological limitations of the organism." In particular, the "limits on computational capacity" of an organism are powerful constraints, requiring that the organism simplify the calculations behind its choices dramatically. "For the first consequence of the principle of bounded rationality is that the intended rationality of an actor requires him to construct a simplified model of the real situation in order to deal with it. He behaves rationally with respect to this model, and such behavior is not even approximately optimal with respect to the real world. To predict his behavior, we must understand the way in which this simplified model is constructed, and its construction will certainly be related to his psychological properties as a perceiving, thinking, and learning animal."[35]

One of the most important of the simplifications that the organism makes in constructing its model of the world, in Simon's view, is the adoption of an extremely simple payoff function. Instead of creating a vast, precise ordering of all possible outcomes, listing each one as marginally better or worse than the next, organisms tend to judge outcomes as either satisfactory or unsatisfactory. The organism either survives or it dies; the game is either won or lost. The chess player, to use Simon's favorite example, does not have to defeat his opponent in the best, most efficient way. He simply has to win. He does not have to choose the best move (or strategy); he just has to choose one of the many potentially winning ones. This simplification drastically reduces the computation necessary for the organism, making it possible for it to conduct simple tests of possible actions and to choose the first acceptable option.

Learning more efficient, more nearly optimal strategies, in this model, could be accounted for by a simple mechanism that raised or lowered the aspiration level of the organism, redefining what was satisfactory according to the situation. Such a higher-level aspiration-adjusting mechanism could be understood as a kind of feedback or control mechanism. In combination with a simple selection mechanism, the resulting system would have the properties of an Ashby-style "ultrastable" system.[36]

Simon believed that this simple model of rational choice was the key to a synthesis of choice and control. "The paradox vanishes, and the outlines of theory begin to emerge when we substitute for 'economic man' or 'administrative man' a choosing organism of limited knowledge and ability."[37] In "Rational Choice and the Structure of the Environment," Simon moved to a description of the nature of the choice mechanisms—and the nature of the environment—that would enable such an organism to survive.[38]

Simon begins "Rational Choice" by asking the reader to consider an extremely simple organism with only a single need (food) and only three modes of activity (resting, exploring, and food getting). The organism can travel over the "bare surface" dotted with "little heaps of food" that constitutes its "life space." This organism has a certain limited perceptual capacity that allows it to see a certain number of "moves" (rooms) ahead along the "branching system of paths, like a maze," that make up its mental world. In addition, this organism's "needs are not insatiable, and it hence it does not need to balance marginal increments of satisfaction," a quality that distinguishes it further from "economic man." Given such a simple creature, the "problem of rational choice" is correspondingly simple: how to "choose its path in a way so that it will not starve."[39]

Simon builds complexity into this simple model step by step, first adding choice mechanisms for allocating time toward the achievement of multiple goals. He then adds the ability to detect "clues" in the environment that lead the organism to food more rapidly and aspiration-adjusting mechanisms that allow it to adapt to the richness or barrenness of the environment. Even with these additions, the result is still a remarkably simple model of rational choice, one more appropriate to "rat psychology" than to human psychology. Nevertheless, Simon believed, "Economics and administrative theory both need models of rational choice that provide a less God-like and more rat-like picture of the chooser. The assumptions of omniscience and intelligence implicit in the utility function model are truly fantastic."[40]

This redefinition of choice also entailed a redefinition of rationality and of the scope of a theory of rational choice. As he wrote in *Models of Man*, "The central task of these essays, then, is not to substitute the irrational for the rational in the explanation of human behavior but to reconstruct the theory of the rational." If this task were accomplished, he believed, then a "return swing of the pendulum will begin," and "we will begin to interpret as rational and reasonable many facets of human behavior that we now explain in terms of affect."[41] Simon later would expand the sphere of rationality to encompass (nearly) all mental activity, including the seemingly ineffable acts of inspiration associated with the highest flights of creativity and the deepest springs of emotion.[42]

Stress, Cybernetics, and Homo Adaptivus

Simon's boundedly rational man was an adaptive machine. In developing this concept, Simon drew on recent work in cybernetics and physiology, particularly that of psychiatrist and neurophysiologist W. Ross Ashby, whose 1952 book *Design for a Brain* had fired Simon's imagination. In keeping with our magical year 1956 theme, it should be no surprise that 1956 also saw the publication of the first textbook in cybernetics, Ashby's *Introduction to Cybernetics*, as well as a reissue (with a slightly revised title) of Alfred Lotka's proto-cybernetic *Elements of Mathematical Biology*.[43] While Norbert Wiener had coined the term *cybernetics* and had introduced its notions of feedback, communication, and control to a wide audience, many high modernists (especially those on the biological side) drew more from Ashby than Wiener (with electrical and communications engineers following the work of Claude Shannon, Harold Hazen, and other communications engineers at Bell Labs).[44] As a result, the publication of Ashby's *Introduction* was eagerly anticipated.[45]

Ashby linked cybernetic concepts, especially feedback, to physiological concepts, especially homeostasis. *Homeostasis* was a term coined by physiologist

and biochemist L. J. Henderson and later elaborated into one of the fundamental concepts of physiology by Walter Cannon in his extraordinarily influential *Wisdom of the Body* (1932).[46] In the hands of mathematical biologists, such as Henderson, Cannon, Lotka, Ashby, Ralph Gerard, Kurt Lewin,
and Anatol Rapoport, homeostasis became a unifying principle, a conceptual
touchstone that framed the organism as an adaptation machine, one forced by
a changing world to adapt continuously. It was so crucial an idea, for some,
that it served as the basis for redefining life and death—life ends when the
ability to adapt ceases—a definition that, once again, blurred the line between
organism and machine.

The concept of homeostasis also led to the formulation of another fundamental concept in postwar human science and biosocial thought, not to
mention postwar popular culture: the concept of *stress*, popularized by physiologist Hans Selye in 1956 (of course) in his best-selling book *The Stress of Life*.
As with all the key works published between 1955 and 1957, the original concept of stress appeared a few years earlier, in a 1950 article by Selye, "Stress and
the General Adaptation Syndrome." In this article, Selye laid out a prestigious
lineage for the concepts of stress and the general adaptation syndrome (his
term for the universal pattern of physiological responses to stress), including
Claude Bernard's concept of the *milieu interieur* and Cannon's interpretation
of homeostasis—and presented his own concept of the general adaptation syndrome as the "keynote" principle for unification of biology: "The keynote of this
unification was the tenet that all living organisms can respond to stress as such,
and that in this respect the basic reaction pattern is always the same, irrespective
of the agent used to produce stress. We called this response the general adaptation syndrome, and its derailments the diseases of adaptation."[47]

In short, "Anything that causes stress endangers life, unless it is met by
adequate adaptive responses; conversely, anything that endangers life causes
stress and adaptive responses." Thus, to Selye, "Adaptability and resistance to
stress are fundamental prerequisites for life." But that capacity to adapt and resist stress, like the cognitive capacity of homo adaptivus, is a "finite quantity."[48]
Thus, much of life—anatomy, physiology, and behavior—is to be understood
as a product of our limited, but still significant, ability to adapt.

The Stress of Life carried this argument further, extending, elaborating, and
teaching the concept of stress to a rising generation of scholars and a broad
literate public. As Selye puts it in his introduction, "No one can live without
experiencing some degree of stress all the time." Indeed, "life is largely a pro-

cess of adaptation to the circumstances in which we exist . . . The secret of health and happiness lies in successful adjustment to the ever-changing conditions on this globe; the penalties for failure in the great process of adaptation are disease and unhappiness." This adaptive process, like all core processes, took place on multiple levels: "The evolution through endless centuries from the simplest forms of life to complex human beings was the greatest adaptive adventure on earth . . . But there is another type of evolution which takes place in every person during his own lifetime from birth to death: this is adaptation to the stresses and strains of everyday experience."[49]

But what is stress? Clearly, the concept of stress is an abstraction, but "concepts always work through abstractions, for it is only by abstracting from the distinct, individual features of each factual object that we can arrive at some common hold on many of them" (49). Selye quotes James Conant several times on this topic: "The history of science demonstrates beyond a doubt that the really revolutionary and significant advances come not from empiricism but from new theories" (194).

The key to this abstraction's utility is to realize that "stress is a condition, a state, and, although as such it is imponderable, that it manifests itself by measurable changes in the organs of the body" (49). It is manifested by a patterned sequence of responses—alarm, resistance, exhaustion—a specific syndrome triggered by an infinite array of stimuli. Thus, the general adaptation syndrome that manifests stress in the body is the logical, natural response of a limited, finite creature facing an infinite array of threats.

The keys to this adaptive response system are the nervous system and the endocrine system. These complex, hierarchic systems, are, to Selye, best understood as (inevitably) communication systems. Hormones are "chemical messengers," of course, and the nervous system is like the telephone or telegraph system in having specific cellular "wires" or "cables" to carry and channel messages. Selye notes that the endocrine system is "more comparable with radio or television networks, in which the emitting station sends its waves indiscriminately in every direction, but only certain sets are tuned to a particular wavelength" (77).

As *The Stress of Life* progresses, Selye seeks the fundamental unit of life, which he sees as the "smallest particle of life which can still respond selectively to stimulation" (213). This fundamental unit, which he terms the "reacton," leads him to a mechanized holism (or holistic mechanism): "All manifestations of life in health and disease are viewed as simple combinations and permutations of individual yes-or-no responses in these ultimate units of life, the reac-tons" (81). Life itself is a binary selection-adaptation machine.

But the goal is not merely to reduce things to binary fundamental units. It is to see how they are combined. To this end, Selye carries the analogy further: "The impression of virtually any color, shape, or movement can be created on an illuminated panel by turning off and on different combinations of colored light-bulbs, though each is capable only of one kind of response. As far as we can see, the human body represents an essentially similar, though enormously more complex, three-dimensional panel" (213). "We hope to show that reac-tons, not cells, are the elementary 'keys' of living matter, and that all the manifestations of normal and pathological life depend only on when, where, and how much (or how many of) these biologic elements are stressed . . . This concept is closely related to that of the German Gestalt school of psychology. Gestalt means literally 'form' or 'shape,' and is used in this sense for a config-uration of separate structures or systems . . . so integrated into a patterns as to constitute a functional unit" (219). Units are thus defined by functions, which may not correspond to specific structural units. Such considerations leave open the possibility (unstated by Selye but quickly grasped by others) that some of those physical structures might be within the organism and others *without*: if the key is communication, if the key is enabling action among dif-ferent "reactons" so that they function as a unit, then a system of people, or of people interacting with machines, can and should be seen as a single unit—a single adaptive "organism."

Selye abstracts the notion of the reacton as functional unit into a concept of the reacton as "a focus of interaction," not a physical thing but "a func-tional plan or pattern which governs the organization of matter" (234). Such a plan is almost alive; indeed, it almost *is* life: "A plan possesses, to an exquisite degree, such accepted characteristics of life as the ability to grow, to reproduce its own kind, and to adapt itself to changing requirements" (234).

Though a physiologist, Selye, like his hero Cannon, does not shy away from spelling out the social implications of stress, homeostasis, and adaptive plans. "Stress is usually the outcome of a struggle for self-preservation (the homeo-stasis) of parts within a whole. This is true of individual cells within man, of man within society, and of individual species within the whole animate world" (253). Thus, one might expect analogous "social adaptation syndromes" in organizations (or even nations), and one surely could hope to find "plans" governing functional responses to stress in firms as well as in bodies. As Selye goes on (at length) to explain, these concepts of stress and homeostasis also lead to a relational, reciprocal morality: to him, the pursuit of gratitude from others is the key to long-term happiness. This produces a complex world indeed, one in which we continually adjust ourselves according to our changing expectations of reciprocal reactions from others, all of whom are continually adjusting their own reactions to us in like manner.

This might seem a stultifying world of cybernetic conformity, but Selye does not see it that way. The very limits of homo adaptivus are the keys to its power, and the limits imposed by society can be the keys to freedom, so long as they do not force the system from a free to an unfree equilibrium. Herbert Simon saw reason as only being possible in man *because* it was bounded. Selye likewise held that "it is only in the heat of stress that individuality can be perfectly molded . . . As I see it, man's ultimate aim is to express himself as fully as possible, according to his own lights" (277). To this end, "the goal is certainly not to avoid stress. Stress is part of life" (299). Rather, the goal is to manage those stressors more intelligently, enabling both group cohesion and individual identity—and, perhaps, just perhaps, even "extending our lives dramatically" (303).

To Selye, this cybernetic utopia of well-adjusted individuals living Methuselean lives lay on the "road ahead." Not just around the corner, but not too far either. Getting there was "largely a matter of time and money," and of course, "for this, organized research teamwork is necessary" (303).

Organisms and Organizations

Not all roads led to cyber-utopia. But the new model of homo adaptivus did promise to recast knowledge of organizations as well as that of organisms. Two recastings of organizations are the revision of the firm seen in C. West Churchman's foundational *Introduction to Operations Research* (1957) and Duncan Luce and Howard Raiffa's *Games and Decisions* (also 1957) and a redefinition of the party (and thereby of democracy) in Anthony Downs's "Economic Theory of Political Action in a Democracy" (also 1957). While these pieces might at first seem far removed from Miller, Bruner, Chomsky, Simon, and Selye, they shared the fundamental conception of humans as limited choosers, with the limits to our ability to make rational choices (and thus to adapt to changing circumstances) arising from our limited access to, and ability to process, information. In all these works, the organization—whether it be the firm or the party—simplifies the world for its members, making rational action possible, though it be rational only within those bounds. As Herbert Simon put it, "The rational individual must be an organized and institutionalized individual."[50] The rational individual did not—could not—exist alone.

Operations Research and the Cybernetic Organization

In *Introduction to Operations Research*, one of the first textbooks on the subject, Churchman needed to begin by defining his field and his basic terms. So what was OR and whence did it come? Churchman answers his own questions by referring to an ongoing trend in business—the "differentiation and segmentation of the management function" in large-scale business organizations, a process that created a new class not only of professional managers but also of managerial problems, which he calls "executive-type" problems. These problems are "a direct consequence of the functional division of labor in an enterprise, a division which results in *organized* activity." In addition to this trend in business (the continuing organizational revolution), Churchman cites the specific impetus of the demands of managing the vast enterprise of war: "During World War II, military management called on scientists in large numbers to assist in solving strategic and tactical problems. Many of these problems were what we have called executive-type problems. Scientists from different disciplines were organized into teams that were addressed initially to optimizing the use of resources. These were the first O.R. Teams."[51]

What distinguished the work of these teams was their "application of scientific methods, techniques, and tools" to these new problems. In particular, OR strove for a "comprehensive" approach to managerial problems, for such problems involved many variables and required knowledge drawn from many fields. The comprehensiveness of OR thus was grounded in a view from the top of the organizational hierarchy, a team-based approach to problem solving, and (equally fundamentally), a "systems approach," as "system implies an interconnected complex of functionally related components. Thus, a business organization is a social or man-machine system" (7).

Churchman does not devote much space to the first of these characteristics (the view from the top), other than to make the usual high modernist's comment that problems at different levels of the organizational hierarchy may look different but are solved in the same way, with the same basic methods. He does, however, go on at great length in regard to the team approach and the systems perspective. To him, the two points went together—OR was the team approach to the analysis of systems: "Another important advantage of the team approach lies in the fact that most man-machine systems have physical, biological, psychological, sociological, economic, and engineering aspects. These phases of the system can best be understood and analyzed by those trained in the appropriate fields. Those in control of a system may be unaware of one of more of these aspects and hence have an incomplete picture of their system. That is, to see a system as a whole means not only see all its components and their interrelationships but also all aspects of its operations" (10–11). An example of this picture in the book is a "flow diagram of repair parts order processing" clearly intended to evoke diagrams of electronic circuitry (40–41). An organism is an organism is a machine, all defined by how they process information.

After outlining the benefits of and goals for OR, Churchman presents his conceptual foundation in chapter 4, "Analysis of the Organization." The chapter is heavily indebted to the cybernetic and cognitive literature discussed above, with multiple citations to Walter Cannon, Warder Allee, Freed Bales, Leon Festinger, and Fred Barrett on the social implications of cybernetics; Alex Bavelas, Fred Chapple, and Walter Cannon on homeostasis; Karl Deustsch (nine different works) on cybernetic communication and control systems in the polity; Donald Hebb, George Homans, and Karl Lashley on serial order in information processing; Harold Lasswell on the structure and function of communication in society; Elton Mayo and Adolph Meyer on the heuristic value of scientific models; George Miller on language and the psychology of communication; Oskar Morgenstern on the theory of organization; and Walter Pitts and Warren McCullogh on how we know universals. In short, it is a "who's who" list of theorists of homo adaptivus.

And how does Churchman put together this hybrid of already hybridized ideas? The key is that they all help the OR team construct a "communication model" of the "organism" (that is, the business organization):

During the late 1930s and the 1940s groups of physiologists, electrical engineers, mathematicians, and social scientists began to work on organizational problems. Many organizations, they found, had similar characteristics. For example, human beings seemed to suffer many faults in their nervous systems which were analogous to faults appearing in electric gun-control mechanisms. Diagrams . . . which biologists and physiologists had drawn of the human nervous system even looked like electric circuit diagrams . . . Groups of such scientists, working in Cambridge, Massachusetts, and elsewhere, soon saw the possibility of developing a generalized organization or control theory that would cut across scientific disciplines. (69–70)

The key to such a theory was to define control in terms of communication, and vice-versa. The first to do so, Churchman states, was Norbert Wiener, who showed that "communication (or information transfer) and control were essential processes in the functioning of an organization. Professor Wiener used information as a general concept, meaning any sign or signal which the organization could employ for the direction of its activities. The information might be an electric impulse, a chemical reaction, or a written message; very generally, anything by which an organization could guide or control its operation" (69).

So conceived, a communication theory would have something of value to say about organisms—systems—of all sizes, shapes, and scales. "Thus, the view of Cybernetics is that (a) Organizations composed of cells in an organism, (b) Organizations composed of machines in an automatic factory or electric communications network, and (c) Organizations of human beings in social groups all follow the essential processes of communication and control in their operations" (69, 71). Thus, "one can often analyze industrial or military organizations, even though they are complex, in the same communication and control terms. Such analysis can be directed toward the construction of a *communications (or control) model* of the organization" (71).

"A communication model is not mathematical," Churchman argued. "It is not used for accurate predictions or calculations. It generally takes the form of a diagram. Such a diagram enables one to bring together, from various fields of research, knowledge about organizations" (71) Indeed, "The communications model can be thought of as a glorified kind of fish net, spider's web, or network of nerves through which 'information' passes or flows" (72). In keeping with the high modern emphasis on models as vital simplifying devices, Churchman also points out that "a model is a miniature of, or compact representation of, an original. Usually models represent relevant points of interest in the original; these points can be combined so that the structure of the model and that of the original are similar" (72).

The particular model is a communication model. Here is one of the great novelties of the new model of humans, machines, and organizations: communication defines the organism, draws its bounds, and connects it not only internally but also (through different channels and different modes) to the outside world. "An organization can be thought of as a group of elements (divisions in a company, operating units in a machine, people in a social group) which are in some way tied together through their communication with each other" (74). Therefore, "the first thing to be determined about an organization is the existing structure of the communication network," a structure that must be mapped before the organization can be understood and its problems solved. That map—the communications diagram—will, of course, "look—on paper—like a road map or circuit diagram" (74).

"Organizations—companies, groups of parts in a machine, the functional elements of the human body—operate together in a communication network, but they also exhibit another characteristic: the elements of an organization operate together to reach or maintain an external goal (or its goal-image within the organization)."[52] Thus, organizations are not only communications systems, they are *teleological* (purpose-driven) systems, whose basic mechanisms are error-controlled feedback mechanisms.

On this basic foundation—an organization is an organism is an adaptive, error-controlled system defined and bounded by information flows that acts in relation to an internally stored image or model of the external world—Churchman then builds a model of the organization as an organism as a computing device. He begins this process of construction much as Selye did, with the simplest possible units. These units, in combination, produce behaviors of startling complexity. Instead of Selye's singular binary "reacton," however, Churchman posits three different fundamental units: a "simple transformation unit" with no goals of its own that only performs operation A on input B to produce output C; a "simple sorting system," which takes an input and sorts it into one of several categories according to built-in rules (reminiscent of *A Study of Thinking*); and a "control unit," which monitors its operations against an external goal (76). Organizations of any and all sizes are but combinations and permutations of these three basic transformation, sorting, and control units.

But the model is not yet complete. "If an organization has several alternatives prepared for action, and also has the rules set up for applying one or the other of them *when external conditions change* (i.e., it can *predict* the best alternative for changing conditions), it can control its own activities more effectively than can a simple feedback system. Such action requires a second-order feedback and implies that a reserve or memory of possible alternatives exists within the organization" (81). There follows a diagram of a feedback circuit with a memory device. "An example of this type of organization—which can switch its standards for different courses of action—is the telephone exchange" (81).

> If an organization can control itself, particularly if it can change its goals, we call it an *autonomous* organization. The autonomy of the automatic goal-changing organization lies in its memory and ability to recall . . . The storing up of information, which allows the organization to prepare various alternatives for action, is a process of *learning* . . . The learning organization's structure changes with time . . . If an organization can collect information, store it in a memory, and then reflect upon or examine the contents of the memory for the purpose of formulating new courses of action, it will have reached a new level of autonomy. The mechanism that considers various goals and courses of action can be called the *consciousness* of the organization (82–83).

Churchman continues, building his machine-organization ever closer to a mind:

> Conscious learning can be selective and take, from a wide range of external information sources, that information relevant to the organization's survival or other major goals. The consciousness may redirect the attention of the organization; make it aware of some happenings and unmindful of others. It can initiate or cease courses of action, based on incoming information; investigate network conditions in the organization; search the organization's memory; and pick up deviations between various actions and the goals which direct them . . . By taking such actions, the organization with a consciousness can direct its own growth (83).

The OR team is just such an organization, just such a consciousness, and it can transform other organizations into conscious creatures capable of deliberate design and forethought.

Churchman, like other high modernists, thus saw both the OR team and the organization it advised as adaptive, goal-directed, information-processing organisms that strove to maintain a dynamic equilibrium with the world outside. Such a view was grounded, in fundamental ways, in the presumption that individual decisionmakers were limited in their ability to gather and process all the relevant information: hence the need for teams of specialists and for simplifying models to orient action. Churchman, however, hoped OR's methods would yield nearly optimal results; though limited when working solo, homo adaptivus could break those limits when organized.

Political Adaptations

This organizational advantage might be conceivable from the narrow view from the top of the firm where one goal (profit) could be presumed to order all others. But what about other organizations—polities or parties—whose goals were not so easily specified or as widely shared? These questions received mixed answers from the high modernists who tackled them at this time: Robert Dahl, in *Preface to Democratic Theory*; Anthony Downs, in "An Economic Theory of Political Action in a Democracy"; and Duncan Luce and Howard Raiffa, in *Games and Decisions*. Of these three works, Dahl's *Preface* is probably the most famous, *Games and Decisions* is probably the most influential (it was the text from which most social scientists learned game theory during the 1960s), and Downs's "Economic Theory" is probably the most revealing of the underlying assumptions of its day.

Dahl's *Preface* is a product of its time as well, for it strives to translate the not-formally-articulated theories of Madisonian and populist democracy into relatively formal systems of definitions and hypotheses. Dahl is aware that some crucial terms cannot be given precise definition and that this is not always a bad thing. But his goal is clarity through formalization, and the result is an endorsement of the realities of democratic politics—a politics in which "minorities rule" in a "polyarchal system" of *many* diverse minorities competing to influence policy through elections, rather than a celebration of an idealized democratic rationality or morality, let alone an ideal "public interest."

Indeed, to Dahl, what we ordinarily describe as democratic "politics" is merely the chaff. It is the surface froth representing only superficial conflicts. "Prior to politics, beneath it, enveloping it, restricting it, conditioning it, is the underlying consensus on policy that usually exists in the society among a predominant portion of the politically active members." Again, such limits are both necessary and good: "Without such consensus, no democratic system could long survive." For example, "it is much more plausible to suppose that the constitution has remained because our society is essentially democratic than to believe that this country has remained democratic because of its constitution."[53]

Despite this affinity with the high modern style of analysis and its emphasis on the necessary limits that structure politics at the most basic levels, Dahl's *Preface* does not draw specific connections between his models of democratic politics and democratic man. Luce and Raiffa, however, seek to bridge the conceptual divide between organization and individual: "The distinction between an individual and a group is not a biological-social one but simply a functional one. Any decision-maker—a single organism or an organization—which can be thought of as having a unitary interest motivating its decisions can be treated as an individual."[54]

Luce and Raiffa frame their analysis of game theory in terms of understanding decisionmaking under conditions of *certainty* (X leads to Y), *risk* (X leads to a known probability distribution), and *uncertainty* (X leads to certain outcomes whose probabilities are unknown). That is, their exposition is structured around decisionmaking under conditions of increasingly incomplete and expensive information. Much of the value of game theory, then, lies in making problems simpler to aid decisionmaking: simple payoff functions, discrete or canalized choices, the collapse of sequences of choices over time into a single choice in the present regarding the strategy to employ, the restriction of an infinite range of choices to a sharply delimited menu of identifiable moves, and so on.

While the assumptions of game theory often are described as "heroic," and often seem to demand things of the chooser of which humans are not capable, the game theory of Luce and Raiffa was not designed with a heroically rational human in mind. Rather, it was explicitly presented as a tool to aid an imperfectly informed, boundedly rational, decisionmaker. The "heroism" of game theory, at that time, lay in the hopes for the method, not in the assumptions about actual humans. A generation later, the model method would be imported back into the mind, with every individual becoming, implicitly, a game theorist of high caliber, but that was not usually the case in the game theory of 1957.

Downs explicitly builds his model of party politics from his model of man as homo adaptivus, the limited chooser. In his landmark article, "An Economic Theory of Political Action in a Democracy" (one cited more than 21,000 times, according to Google Scholar as of September 2014), Downs applies an economist's cold calculus to the hot buttons of politics, smashing many an icon of democratic theory in the process (with more than a hint of pride).

Downs begins by observing, "most welfare economists and many public finance theorists implicitly assume that the "proper" function of government is to maximize social welfare." To him, this is the wrong place to begin, especially since Kenneth Arrow (Downs's PhD adviser) had shown in *Social Choice and Individual Values* (1951) that "no rational method of maximizing social

welfare can possibly be found unless strong restrictions are placed on the preference orderings of the individuals in society." For a science (welfare economics) to devote itself to constructing impossible theories is a waste of time and
energy; similarly, it is useless for political science to assume that its key actors
(politicians, parties, and voters) attempt to achieve the impossible in practice.
Rather, Downs argues, a valid theory of democratic politics should borrow
from general economic theory, which assumes that private actors act according
to "their own selfish motives," with the broader economic or political system
resulting from decisions based on these motives, not from decisions based on
altruism. The result of such a new approach would be a new "positive," "political equilibrium theory."[55]

Downs lays out his definitions and axioms, building first a hyper-simplified
model, one intended to be unrealistic in instructive ways. Then, he moves
to build a more robust model by responding to the oversimplified model's
shortcomings.

The logic is swift and forceful; the claims dramatic; the language dry and
matter-of-fact. The axioms begin with the assumption that "each political
party is a team of men who seek office solely in order to enjoy the income,
prestige, and power that go with running the governing apparatus." If the party
wins, he further assumes that "the winning party (or coalition) has complete
control over the government's actions until the next election." (It cannot
eliminate its political rivals by force or fiat, however.) In addition, "government's economic powers are unlimited. It can nationalize everything, hand
everything over to private interests, or strike any balance between these extremes." And, most important, "every agent in the model—whether an individual, a party or a private coalition—behaves rationally at all times; that is, it
proceeds toward its goals with a minimal use of scarce resources and under-
takes only those actions for which marginal return exceeds marginal cost"
(137). If these seem like heroic assumptions, that is precisely the point.

From these definitions and assumptions come his central hypothesis: "Political parties in a democracy formulate policy strictly as a means of gaining
votes. They do not seek to gain office in order to carry out certain preconceived
policies or to serve any particular interest groups; rather they formulate policies
and serve interest groups in order to gain office. Thus their social function—
which is to formulate and carry out policies when in power as the government—
is accomplished as a by-product of their private motive—which is to attain the
income, power, and prestige of being in office" (137). Significantly, "this hy-

pothesis implies that, in a democracy, the government always acts so as to maximize the number of votes it will receive. In effect, it is an entrepreneur selling policies for votes instead of products for money. Furthermore, it must compete for votes with other parties, just as two or more oligopolists compete for sales in a market . . . We cannot assume a priori that this behavior is socially optimal any more than we can assume a priori that a given firm produces the socially optimal output" (137).

Downs next moves to construct a model of government decisionmaking in two situations: the first, where there is perfect knowledge and information is costless, and the second, where knowledge is imperfect and information is costly. Crucially, the basic features of our "real" political system *all* are necessary consequences of a world where political actors are rational but have imperfect knowledge and remedying that imperfection is costly.

The first situation, obviously, is unreal, but Downs (like many high modernists) believed that hypersimplified models of unreal situations could be valuable. These situations could be used to model the basic structure of relationship among elements—in this case, between a government and its people—and their very unreality could point to crucial variables or parameters whose absence produced the unreality. In this case, the basic model of government-people relations is that of the management of a large firm trying to decide what product mix at which prices will bring the greatest return. So analyzed, "the political structure of a democracy can be viewed in terms of a set of simultaneous equations similar to those often used to analyze an economic structure" (138). In addition, "because the citizens of our model democracy are rational, each of them views elections strictly as means of selecting the government most beneficial to him" (138). In other words, narrow self-interest is rational—indeed, it is the definition of rationality, not the enlightened pursuit of an idealized (or imaginary) public interest. Thus, "each party resembles a player in an N-person game or an oligopolist engaged in cut-throat competition" (138).

In this simplified model, every citizen is equal and there is no need for parties or campaigns or opinion leaders, or for any kind of leader for that matter. One simply could run all policy decisions through a vast polling machine and follow the majority on every point. That is obviously not the world we live in, and for Downs, it is not even a desirable ideal, even were it possible. Still, this simple model sets the basic structure of relationships and needs only one major alteration to align with reality. That alteration is that government-individual relationships, like all social or economic relationships (and all human encounters in the world) are structured not just by rational self-interest but also by imperfect information. "Lack of complete information on which to base decisions is a condition so basic to human life that it influences the structure of almost every social institution" (139).

Because of imperfect information, and the costs in time, energy, and money that are required to remedy this imperfection on any given question, the whole structure of the system changes. Some people suddenly become more important than others because they influence decisions by the information they present to them and how they select and present it. "The government, being rational, cannot overlook this fact in designing policy. As a result, equality of franchise no longer assures net equality of influence over government action. In fact, it is irrational for a democratic government to treat its citizens with equal deference in a world in which knowledge is imperfect . . . In other words, lack of information converts democratic government into representative government" (140). Thus, lack of information creates, necessarily and unavoidably, a political hierarchy. To use the language of the cognitive psychologists, political information is "chunked" and "recoded" as it moves up the tree of political organization.

This reasoning implies that a democratic government in a rational world will always be run on a quasi-representative, quasi-decentralized basis, no matter what its formal constitutional structure, as long as communication between the voters and the governors is less than perfect. Another powerful force working in the same direction is the division of labor. To be efficient, a nation must develop specialists in discovering, transmitting, and analyzing popular opinion, just as it develops specialists in everything else. These specialists are the representatives. They exercise more power, and the central planning board exercises less, the less efficient are communication facilities in society . . . [I]mperfect knowledge makes the governing party susceptible to bribery. In order to persuade voters that its policies are good for them, it needs scarce resources, such as television time, money for propaganda, and pay for precinct captains. One way to get such resources is to sell policy favors to those who can pay for them, either by campaign contributions, favorable editorial policies, or direct influence over others. Such favor buyers need not even pose as representatives of the people. They merely exchange their political help for policy favors—a transaction eminently rational for both themselves and the government (140–41).

Thus, "inequality of political influence is a necessary result of imperfect information, given an unequal distribution of wealth and income in society . . . This outcome is not the result of irrationality or dishonesty. On the contrary, lobbying in a democracy is a highly rational response to the lack of perfect information, as is government's submission to the demands of lobbyists" (141). In short, everything "good-government" types praise—such as the nonpartisan pursuit of the public interest—is a fantasy, and everything they lament—hierarchy, influence peddling, lobbying, bribery, and corruption—are "highly rational" adaptations to a world of incomplete information.

Downs then tackles the manifest existence of political ideologies in democratic politics, admitting, "Since the parties in this model have no interest per se in creating any particular type of society, the universal prevalence of ideologies in democratic politics appears to contradict my hypothesis" (141). He quickly shoots down this objection, however: "But this appearance is false. In fact, not only the existence of ideologies, but also many of their particular characteristics, may be deduced from the premise that parties seek office solely for the income, power, and prestige that accompany it. Again, imperfect knowledge is the key factor" (141).

People do not have the time or the information necessary to compare parties on all the issues, and the ramifications of policy choices often are unclear. Ideologies provide useful summaries, focusing attention on key differences, giving a cognitively less demanding heuristic for choice for the voter. If there is a correlation between a party's ideology and its policies, then a voter can vote rationally based on ideology. Changing ideologies is risky because voters only pay attention to ideologies insofar as they are useful guides to actual policy choices, and no party can afford to look unreliable and unpredictable.

Discussing the conditions necessary for political stability, a parallel to the central concern of economists for the stability of economic equilibrium, Downs reaches much the same conclusion as Dahl in the *Preface*: "Stable government in a two-party democracy requires a distribution of voters roughly approximating a normal curve," and "democracy does not lead to effective, stable government when the electorate is polarized . . . If a majority of voters are massed within a narrow range of that scale [from left to right], democratic government is likely to be stable and effective, no matter how many parties exist . . . Thus the distribution of voters—which is itself a variable in the long run—determines whether or not democracy leads to effective government" (143, 145).

An even more surprising inference follows: if the parties resemble each other fairly closely, as they will in a stable system, it is actually *irrational* for the ordinary voter to take the trouble to become better informed. There will be some cost involved in time and energy to gain the information necessary to choose between the parties and voting "correctly" will bring almost no tangible reward to most individuals compared to voting "incorrectly." Therefore, "we reach the startling conclusion that it is irrational for most citizens to acquire political information for purposes of voting." Hence, "ignorance of politics is not a result of unpatriotic apathy; rather it is a highly rational response to the facts of political life in a large democracy" (146).

This results in a tragedy of the commons for political decisionmaking:

> Consequently, it is rational for every individual to minimize his investment in political information, in spite of the fact that most citizens might benefit substantially if the whole electorate were well informed. As a result, democratic political systems are bound to operate at less than maximum efficiency. Government does not serve the interests of the majority as well as it would if they were well informed, but they never become well informed. It is collectively rational, but individually irrational, for them to do so; and, in the absence of any mechanism to insure collective action, individual rationality prevails (148).

Thus another icon of democratic theory falls, with the informed voter cast into the dustbin along with the public interest and the selfless public servant. Yet Downs's essay is not a gloomy "two cheers" for democracy. Rather, it is a strong defense of the necessity—even the virtues—of the messy realities of democratic politics, as it actually exists. As Downs argues, "much of the evidence frequently cited to prove that democratic politics is dominated by irrational (non-logical) forces in fact demonstrates that citizens respond rationally (efficiently) to the exigencies of life in an imperfectly informed world" (149).

To explicate this point—one with which cognitive psychologists such as Miller and Bruner and Simon would have agreed wholeheartedly—Downs returns to the word *rational* and gives it a distinct meaning: "I have employed the word 'rational' instead of its synonym 'efficient' throughout this article because I want to emphasize the fact that an intelligent citizen always carries out any act whose marginal return exceeds its marginal cost. In contrast, he does not always make use of logical thinking, because, under some conditions, the marginal return from thinking logically is smaller than its marginal cost. In other words, it is sometimes rational (efficient) to act irrationally (non-logically), in which case an intelligent man eschews rationality in the traditional sense so as to achieve it in the economic sense" (149).

The bounds of reason, the limits to knowledge, and the costs of information are thus necessary to any theory of "man," just as is the assumption that mind and machine, organism and organization, behave rationally within those bounds. On those two commandments hung all the laws of the high modern prophets.

A System of Systems

Boundaries and limits play different but equally vital roles at the grand scale of the social-cultural system as a whole. Parsons and Smelser's *Economy and Society* demonstrates similarities across scale in high modern work, with their "society" having to solve the same adaptive problems that any cybernetic organism does. Indeed, to Parsons and Smelser, the actor in a social system is just as easily a "collectivity" as an individual.[56] Meanwhile, A. F. C. Wallace's "Revitalization Movements" deals with exactly the area to which Parsons and Smelser devote the least attention: the construction and renewal of the psychic system of the individual within the nested, differentiated system of systems that is the culture. In both these works the organism or organization is an error-controlled adaptive machine, subject to environmental challenges or stresses that threaten its internal equilibrium, but which has developed limited yet powerful homeostatic mechanisms that (usually) produce a satisfactory (if never optimal) result.

Economy, Society, and Cognitive Limits

Though *Economy and Society* is probably the least well remembered of Talcott Parsons's major works today—with *The Social System* and *The Structure of Social Action* holding pride of place for both Parsonians and their foes—Parsons himself considered *Economy and Society* a peak moment, a culminating work in his struggle to redefine the nature and scope of social theory in general and economic theory in particular. It was his most ambitious attempt to sketch a "Columbian map" of the social world, and both its significance as an exemplar of high modernism and its failure as a program for the reform of economic theory followed from its grand ambitions (4). If it had been a bit more a product of its time—or perhaps a bit less—Parsons might have succeeded.

The core argument of *Economy and Society* is simply put: "Economic theory is a special case of the general theory of social systems and thus of the general theory of action" (309). It is such because an economy is "a special type of social system" (310). As a special type, a differentiated subsystem within the larger social system, "the economy, like any social system, exchanges inputs and outputs over its boundaries with its situation" (310). Thus, "a theoretical scheme other than economic theory is the only possible way to analyse these non-economic factors in such a way as to articulate successfully with economic theory" (311). In short, a general theory of the social system is necessary to understand the economy properly.

Without such an understanding, even sophisticated theorists are likely to misdefine their terms, waste effort solving nonsensical problems, and conjure illusory supports from unexamined assumptions about "human nature." For example, Parsons and Smelser state that "to assume any generalized propensity of human nature such as the 'rational pursuit of self-interest' is precluded by our analysis" (306). (*Homo economicus* is dispatched dispassionately, as he deserves.) Instead, economic rationality is found to be an institutionalized value system "appropriate to the economy as a differentiated sub-system of the society; it is institutionalized in the economy and internalized in personalities in their roles as economic agents" (306).

Similarly, Parsons and Smelser recast such fundamental terms as supply, demand, land, labor, capital, entrepreneurial action, rent, production, consumption, wealth, and income in the new language of systems. Wealth, for example, refers to the economic resources that may be mobilized to solve the society's problems of adaptation to its environment. Perhaps most strikingly (and arrogantly, to an economist), these redefinitions lead Parsons and Smelser to dismiss the grand edifice of utility theory and the signal achievement of Kenneth Arrow in *Social Choice and Individual Values* as theoretically meaningless: "Since the development of individual motivation is, in the relevant aspects, conceived as a process of the internalization of social norms and not as the independently given basis of social processes and social values, such issues as the assessment of individual preference lists . . . their relation to community welfare, etc., are theoretically meaningless" (32).

To Parsons and Smelser, the economy is a differentiated subsystem of the larger economy, with its particular role being the solution of the "adaptive" problems of society (as opposed to the "tension management," "integration," and "goal attainment" problems). This definition has implications both for how one is to understand the economy and for who is to study it. Perhaps the most important of these implications is that *production and consumption lie on different sides of the boundary between the economic subsystem and the larger soci-*

ety. The economy produces, the society consumes, and the exchange is a "boundary process" (19–21). Given that the exchange long had been the focus of economic analysis and that the new economics of the postwar period centered on choices, especially the choices of the consumer, this separation of production and consumption was not merely unacceptable but nearly unintelligible to the vast majority of economists. To accept this reconceptualization of the exchange would take consumption out of the hands of economists and give it to sociologists and social psychologists and make the exchange a boundary process fit to be studied only by those with a sufficiently broad perspective to hold both the major subsystems in their compass—systems theorists such as Parsons and Smelser.

This grand theoretical schema had many merits: a society is many things, and one of them is probably a system of functionally differentiated subsystems linked by essential boundary processes. But as an epistle preaching the good news of social systems theory to economists, *Economy and Society* was a flop.

It was a revealing flop, however. As we have seen, it was based on many of the same premises as contemporary paradigm-setting works in other fields and even had some kinship to flourishing aspects of economic theory, such as Keynesian macroeconomics and OR. So why did this ambitious, abstract, synthetic social theory have far more limited appeal than either its own earlier forms (Parsons's *The Social System*) or other equally ambitious (and almost as abstract) works? For example, *Organizations*, by James March, Harold Guetzkow, and Herbert Simon, was published in 1958 to enormous acclaim and has remained on reading lists in schools of business and public administration for over fifty years, though it paints a similar picture of the organization to *Economy and Society*, deals in quite abstract social-psychological theory, and is heavily indebted to earlier Parsonian social theory.[57] Why the difference in response?

The reasons for the relatively cold response to *Economy and Society* in contrast to *Organizations* (at least outside the confines of structural-functional sociology), are fourfold: first, and probably foremost, *Organizations* is well written, and *Economy and Society* is really a chore to struggle through. It isn't artfully unintelligible the way fashionable theory can be; it's just a long slog. But that observation doesn't take us far analytically, other than to provide good advice for future theorists: it always helps to write well.

Second, the mode of presentation is top-down, with many of the key concepts and insights from different fields, especially economics, being presented as *derived from* the author's grand model, rather than the grand model being *built up* from the accomplishments of the various fields. It always helps to build on others rather than to bash them, even when one is reinterpreting their fields for them.

Third—and probably most significant for economists (and choice theorists in other fields)—the aforementioned separation of production and consumption, with production being internal to the economy and consumption being external to it, was deeply destructive to economic theory and disciplinary identity without offering any new, clear solutions to recognized problems within economics.

Last, and most relevant to the story of high modern social science, there is one fundamental aspect of the new model of homo adaptivus that is missing from *Economy and Society*: a focus on the *limits*, especially the cognitive or information-based limits, of the adaptive organism/machine/society. Nowhere do Parsons and Smelser claim that their model human or model society has no limits, of course. But in all the other the primary documents describing homo adaptivus, limits were fundamental to the basic structure of the organism; without limits humans could not think, could not act, could not live and be free. Yes, to Parsons and Smelser, social norms limit actions, but there is little sense of the limits to such norms, and those norms are not presented as adaptations to the fundamental limits to human reason. Similarly, Parsons and Smelser do not investigate the limits to the knowledge or the processing power of the individual, the organization, the institution, or the society.

Without the notion of cognitive limits, and thus of a vital, nontransparent role for communication, social systems theory easily becomes static, for it is imperfection that drives adaptation and change. It loses the essential connection between a model of humans and a model of society that makes for a powerful paradigm. In addition, it loses the empirical anchor that gives theories purchase on reality; without a sense of cognitive and physical limits, the system of linked subsystems is an arbitrary analytic construct, with a structure determined only by the nature of systems in general, rather than one defined by the nature of humans as problem solvers operating within specific bounds.

Revitalization Movements and Paradigm Shifts

One particularly powerful example of how this missing element could make all the difference is A. F. C. Wallace's "Revitalization Movements," one of the most widely cited and most downloaded articles in anthropology of the past sixty years. It is an almost perfect example of high modern social science, one with strong kinship to and influence on modernization theory, social systems theory, and culture-and-personality anthropology. It draws explicitly from "the notions of order and field, function and equilibrium, the organismic analogy, the concept of homeostasis, and certain ideas from cybernetics, learning and perception, and the physiology of stress," all of which are "necessary to justify [the] assumptions on which the revitalization hypothesis is based."[58]

This "revitalization hypothesis" is that a large class of "phenomena of major cultural-system innovation are characterized by a uniform process, for which I propose the term 'revitalization'" (264). More specifically, "a revitalization movement is defined as a deliberate, organized, conscious effort by members of a society to construct a more satisfying culture. Revitalization is thus, from a cultural standpoint, a special kind of culture change phenomenon: the persons involved in the process of revitalization must perceive their culture, or some major areas of it, as a system (whether accurately or not); they must feel that this cultural system is unsatisfactory; and they must innovate not merely discrete items, but a new cultural system, specifying new relationships as well as, in some cases, new traits" (265).

This concept of a revitalization movement is a major innovation, for it creates a new category of culture change.

> The classic processes of culture change (evolution, drift, diffusion, historical change, acculturation) all produce changes in cultures as systems; however, they do not depend on deliberate intent by members of a society, but rather on a gradual chain-reaction effect . . . This process continues for years, generations, centuries, millennia, and its pervasiveness has led many cultural theorists to regard culture change as essentially a slow, chain-like, self-contained procession of superorganic inevitabilities. In revitalization movements, however, A, B, C, D, E, . . . N are shifted into a new *Gestalt* abruptly and simultaneously in intent (265).

The concept of a revitalization movement, Wallace explains, "implies an organismic analogy."

> This analogy is, in fact, an integral part of the concept of revitalization. A human society is here regarded as a definite kind of organism, and its culture is conceived as those patterns of learned behavior which certain "parts" of the social organism or system (individual persons and groups of persons) characteristically display. A corollary of the organismic analogy is the principle of homeostasis: that a society will work, by means of coordinated actions (including "cultural" actions) by all

or some of its parts, to preserve its own integrity by maintaining a minimally fluctuating, life-supporting matrix for its individual members, and will under stress, take emergency measures to preserve the constancy of the matrix. Stress is defined as a condition in which some part, or the whole, of the social organism is threated with more or less serious damage (265).

As is typical of high modernists, Wallace explains that this organismic analogy is no mere analogy. There are homologies and identities across levels of organization and scale, and every larger system is composed of subsystems nested within: "As I am using the organismic analogy, the total system which constitutes a society includes as significant parts not only persons and groups with their respective patterns of behavior but also literally the cells and organs of which the persons are composed. Indeed, one can argue that the system includes nonhuman as well as human subsystems. Stress on one level is stress on all levels" (266).

Continuing with the high modern emphasis on information and communication, Wallace writes, "The holistic view of society as organism integrated from cell to nation depends on the assumption that society, as an organization of living matter, is definable as a network of intercommunication. Events on one subsystem level must affect other subsystems (cellular vis-à-vis institutional, personal vis-à-vis societal) at least as information; in this view, social organization exists to the degree that events in one subsystem are information to other subsystems" (266).

Furthermore, "that regularity of patterned behavior which we call culture depends relatively more on the ability of constituent units autonomously to perceive the system of which they are a part, to receive and transmit information, and to act in accordance with the necessities of the system, than on any all-embracing central administration which stimulates specialized parts to perform their function" (266). Thus, the limits to knowledge preclude such central direction, just as they demand that the individual

maintain a mental image of the society and its culture as well as of his own body and its behavioral regularities, in order to act in ways which reduce stress at all levels of the systems. The person does, in fact, maintain such an image. This mental image I have called "the mazeway," since as a model of the cell-body-personality-nature-culture-society system or field, organized by the individual's own experience, it includes perceptions of both the maze of physical objects of the environment (internal and external, human and nonhuman) and also of the ways in which this maze can be manipulated by the self and others in order to minimize stress. The mazeway is nature, society, culture, personality, and body image, as seen by one person (266).

Such mental images, mazeways, cognitive maps, or models of the world are never perfect. Perfection is not useful. Their goal necessarily is to enable actions that reduce stress to manageable—survivable—levels, not to lead people to either individually or socially optimal choices. The goal is not triumph but homeostasis.

Because neither adaptation nor mental image is ever perfect, because the world around us is never static, stress is inescapable, and any number of stressors can impair a cultural system's (or a mazeway's) "efficiency": changes in climate, flora, or fauna; military defeat; and so on. Similarly, a mazeway may be maladapted for an individual in relation to the larger community that is his or her social environment. The evolution of a species, the adaptation of an individual organism to its environment, and the adaptation of a cultural system to its environment are all parallel developments, with different mechanisms (especially different communications mechanisms) becoming important at different scales.

Even if a mazeway could be perfect in a given moment, which it cannot, it could not be perfect at all times, in all places. Part of the infinitude of the world is temporal; time and change bring stresses and demands to adapt. A sufficiently strong (or healthy) mazeway provides a useful guide for such adaptation to a changing world, but either an insufficiently robust mental image of the world, or a world of changes that are too big or come too fast will impose intolerable stresses. Such stresses may be amplified rather than damped by acting in accordance with the existing mazeway. "Whenever an individual who is under chronic, physiological measurable stress, receives repeated information which indicates that his mazeway does not lead to action which reduces the level of stress, he must choose between maintaining his present mazeway and tolerating the stress, or changing the mazeway in an attempt to reduce the stress. Changing the mazeway involves changing the total *Gestalt* of his image of self, society, and culture, of nature and body, and of ways of action" (266–67). Such a total shift is rare but vital to the evolution of the culture, for it creates radically new cultural forms.

These innovations are individual cultural mutations; a particular individual, in Wallace's account, has a transformative dream or visionlike experience that shows the way to a new Gestalt—to a new way of life and a new way of seeing the world and one's place in it. If the stresses that lead one individual to such a revelation are felt widely enough, and if the vision can be communicated to others who feel a similar need for a new mazeway, then the result is not a dream but a movement.

The revitalization movements that emerge in response to chronic, severe stresses are enormously important to Wallace. He argues that "all organized religions are relics of old revitalization movements, surviving in routinized form in stabilized cultures, and that religious phenomena per se originated . . . in the revitalization process—i.e., in visions of a new way of life by individuals under extreme stress" (268). Thus, at a fundamental level, crucial social changes come as a result of people feeling a need to make better sense of their world in order to take effective action in it; reform is a problem of cognition as well as action.

It is a problem, and a solution, that occurs on many scales, in many times and places: a conceptual scheme, while never optimal, once had been satisfactory, but changes in the world, to which it could not be adapted successfully, eventually tipped the balance from "good enough" to unsatisfactory, and, in a "Gestalt switch," a new conceptual scheme—or paradigm—(with a new exemplary individual to model one's life after) sweeps the field.

The parallels to a Kuhnian paradigm shift are not accidental, and they grow stronger the farther one pushes. In earlier work, Wallace described a culture as "those ways of behavior or *techniques of solving problems* which . . . have a high probability of use by . . . society."[59] In addition, just as scientific revolutions have a structure in time for Kuhn, so too do revitalization movements for Wallace.

Wallace describes his approach as "event analysis"—the study of processes that unfold over time ("diachronic sequences"), which can be analyzed as units—just as game theory combines a sequence of moves linked by a common logic into a player's *strategy*, just as Bruner's learning theory combines many individual acts of perception and selection into identifiable sequences or processes of concept *attainment* (268). Wallace further postulates that "events or happenings of various types have genotypical structures independent of local cultural differences: for example, the sequence of happenings following a severe physical disaster in cities in Japan, the United States, and Germany, will display a uniform pattern, colored but not obscured by local differences in culture" (268).

Returning to the specific case of revitalization movements, Wallace then delineates the "structure of the revitalization process," which comprises five "somewhat overlapping" stages: "(1) Steady State; (2) Period of Individual Stress; (3) Period of Cultural Distortion; (4) Period of Revitalization (in which occur the functions of mazeway reformulation, communication, organization, adaptation, cultural transformation, and routinization), and finally, (5) New Steady State" (268). To translate into the Kuhnian terms that soon became an academic lingua franca, these stages might become (1) normal science, (2) increasing awareness of anomalies, (3) breakdown of the old paradigm in the face of anomalies that cannot be avoided (4) period of revolution in which a new paradigm and its exemplars are articulated, and (5) new normal science.

As is common with high modern social science, this analogy between cultural revitalization and scientific revolution is no mere analogy but a direct connection. Kuhn was trained in history of science—and psychology—by the leading lights of postwar high modernism, including James Conant, Jerome Bruner, and George Miller, and he produced his first great work, *The Copernican Revolution*, in 1957, during the takeoff of high modernism. Central to his work was the concept of a "conceptual scheme" in science, an idea inspired by Conant (and Henderson and Whitehead, among others). The conceptual scheme derives from observations yet "transcends" them. Even more, it performs certain invaluable functions for those who use them, and the "evolution of any scientific conceptual scheme . . . depends upon the way in which it performs these functions."[60] And what are those functions?

"Perhaps the most striking characteristic . . . is the assistance that it gives the memory. This characteristic of a conceptual scheme is often called conceptual economy." Without such a scheme, the world (e.g., the positions and movements of the stars) is a welter of factoids, impossibly numerous and infinitely complex. With such a scheme, the "model replaces the list," and the universe is simplified and rendered intelligible, as patterns now can be discerned amongst the blooming, buzzing confusion of the world.[61]

A good conceptual scheme has meaning for those who believe it. To them, the scheme is true in a deep sense, reflecting the fundamental reality of things. Such a commitment is "always rash," for the "history of science is cluttered with the relics of conceptual schemes that were once fervently believed." Indeed, "there is no way of proving that a conceptual scheme is final." Still, "rash or not, this commitment to a conceptual scheme . . . seems an indispensable one, because it endows conceptual schemes with one new and all-important function. Conceptual schemes are comprehensive; their consequences are not limited to what is already known." Thus, a conceptual scheme's functional requisites are that it be *economical* (and simplify the world in an intelligible way), that it be *satisfying* as an account of the true state of things (and not just as a means of "saving the appearances" or as a product of clever artifice), and that it be *fruitful* (and thus a predictive guide to action and experience).[62]

There is also a "logical structure" to scientific revolutions that unfolds over time.[63] Similarly, a logical structure unfolds over time to the cultural revolutions that Wallace terms revitalization movements—and to the economic revolutions associated with industrial modernization, as Walt Rostow argued in "The Take-Off," published, of course, in 1956. In this cornerstone of modernization theory, Rostow describes the stages of economic growth, the stresses that accumulate in a traditional economy (the old "pre-paradigmatic" economy, one might say) as it struggles to adapt to the rapid changes of the world around it, and the economic paradigm shift to industrialism and sustained growth.[64]

This high modern understanding of the process of change—of logical sequences that are uniform enough that a sequence of events over time can be seen as a unit (as a "general adaptation syndrome," perhaps?)—was grounded in a conceptual scheme of nested (not chained) systems, of information flows and energy conversions, of stresses and adaptations, and, crucially, of humans as beings with significant but limited cognitive powers, beings who necessarily construct simplified mental models of the world to manage their stresses and take action in a satisfactory, if never optimal, way.

In this view, the genius of modern science (and of many modern institutions, such as market economies or large-scale organizations guided by the principles of OR) was that its structure and norms enabled a more comprehensive, coherent model of the world to be built, maintained, tested, and taught than any individual ever could create alone. Though no paradigm, model, metaphor, world picture, or conceptual scheme ever could be perfectly rational according to an Olympian standard, a field with a paradigm to orient it could be progressive, cumulative, and powerful—which is a fair description of high modern social science itself.

Conclusion

Homo adaptivus (a complex, error-controlled, hierarchically structured, functionally differentiated, system of nested systems whose structures, functions, and behaviors are fundamentally shaped by its cognitive limits) has assumed many guises: a cybernetic problem solver and category maker, a homeostatic stress manager, an economic and political decisionmaker, a creator of new cultural gestalts, a traveler in a mental mazeway, and a model-building conceptual schemer, thinker, tinker, and science maker. The different disciplinary clothes that homo adaptivus wore did matter, but in 1956–57, beneath those clothes walked much the same man—for those who embraced high modernism, at least.

Why this unity amid the diversity of postwar social science? And why in 1956–57? The answers to these questions are closely related.

The general question of unity amid diversity is best answered on a similarly general level. As argued in the introduction and chapter 1, the primary context in which to understand high modern social science is the continuing organizational revolution, which posed new problems and set new questions, involved new patrons with new resources and new goals, and emphasized new organizational technologies (especially information and communications technologies) as tools, models, and metaphors.

More specifically, the experience of being part of a world at war (in World War II and the Cold War) taught two generations of social scientists the virtues of interdisciplinary research teams, inspired them with the conviction that their ideas could matter directly in the world, and impressed upon them the importance of communication, command, and control in large enterprises. The grand scale of the social challenges facing the world—economic depression, total war across the globe, the clash of superpowers, the collapse of world-spanning empires, and the sudden rush to modernity (whatever that might prove to be) of more than half the world's population—worked against local explanations. One man's unemployment might be due to his inherent flaws; tens of millions losing their jobs overnight could not be due to their faults as individuals but rather to the faults of the system (a point Paul Samuelson made to great effect in his textbook *Economics, an Introductory Analysis*, first published in 1948 but revised heavily and marketed even more successfully in 1957).

Similarly, one did not win battles in World War II or the Cold War through individual heroism, skill, or puissance at arms. One won through system: logistics, training, communication, mass production, and systematic innovation through research and development. The greatest general was the one who built, managed, and mastered the best production system for war power (Eisenhower), not the most inspired battlefield commander (Patton).

The new patrons of postwar social science shared these commitments and goals, and they funded programs aimed at producing knowledge and solving problems of a certain kind. In addition, technological advances in organizational technologies provided new tools to think with, from radio and television and their associated networks to feedback controlled servomechanisms (frequently used in weapons systems and in electronic circuitry), to the first digital computers, which had an enormous inspirational effect on experimental psychologists, among others, especially once those computers were able to store programs.

Thus, by the early 1950s, there had gathered a confluence of men, money, motivation, and machinery. The early 1950s, not surprisingly, saw breakthrough innovations in tools and concepts, usually developed as solutions to specific problems. By the mid-1950s, the success of these solutions to problems both old and new meant that the time was ripe for reflection, generalization, synthesis—and revolution.

Amid all this emphasis on system and structure and the malleability of the individual, however, the longstanding American reverence for the agency of the individual could not and would not be discarded. It was the totalizing nature of Nazism and Soviet communism that made those ideologies so evil, after all. What was needed was a model of man that acknowledged the reality of external influences and constraints while still saving a place for the individual as a rational and independent agent. So against the extremes of free choice and total control, high modernists charted what they saw as a middle way: the way of bounded rationality, of limited choice, of satisfactory adaptation to the inevitable stress of life.

The high modernists' solution to the dangers posed by maladaptation and the bounds to reason was to construct systems to produce that rationality. Just as science's institutions and norms enabled a greater scope for reason and progress, so too could the firm's or the nation's. Thus, to empower the individual problem solver, the *system* needed to be improved—and empowered. This logic, however, could lead to both triumph and disaster.

CHAPTER FOUR

Producing Reason

[We] need models of rational choice that provide a less God-like and more rat-like picture of the chooser.

Herbert Simon, 1954

When Herbert Simon wrote to psychologist Ward Edwards in 1954, Edwards had just published an impressive survey article in the *Psychological Bulletin*, "Theories of Decision Making," which introduced many psychologists to the explosion of postwar work on decisions and choices, especially game theory, utility theory, and statistical decision theory.[1] Edwards's bibliography for this article ran to over 200 items, over 90 percent of which had been published in the ten years since von Neumann and Morgenstern's monumental *Theory of Games and Economic Behavior*.[2]

While Edwards and Simon differed somewhat in their estimation of the virtues of recent theories of choice, Edwards's review article and Simon's response both reflected a perspective on humans and their behaviors that was quite new to, yet startlingly widespread in, postwar social science. In this view, we humans are limited creatures—animals, not gods—and we are, fundamentally, choosers. We are not defined by our bodies, our souls, our experiences, our hopes, or our dreams. We are defined by our choices.

The limited chooser, the bounded selector, the finite problem solver: that is an interesting and important pairing of ideas.[3] When these two ideas came together in American social science after World War II, the combination triggered a boom in "decision science" and a change in the meaning of choice.

In recent years, there have been several fine studies of the many different pieces of the larger puzzle that were the postwar sciences of choice. These studies, generally speaking, have analyzed operations research, game theory, systems analysis, cost-benefit analysis, utility analysis, statistical decision theory, and all the varied sciences of choice in relation to two basic contexts; they have examined the institutions in which researchers carried out their work (RAND, Case Western Reserve University, Carnegie Institute of Technology, the Cowles Commission), and the new network of patrons of Cold War social science.[4]

While several of these studies have been nothing less than brilliant, there is another crucial context, a third theme that must be heard before the many variations in the sciences of choice can be harmonized. That context is the ongoing debate within the social sciences and in the broader political culture regarding the role of reason in human affairs.

To put it plainly, social scientific discourse about choice from the 1920s to the mid-1970s was part of a discourse about reason and the prospects of democracy. The social scientists who embraced the sciences of choice saw in those fields a new answer to an old question: are humans governed by reason, and if they are not, can they govern themselves?

The answer proposed by many postwar researchers was a blend of pessimism about the scope and quality of human reason and optimism about the power of social and technical mechanisms for producing rational choices. Herbert Simon captured this sentiment well when he stated, "The rational individual is, and must be, an organized and institutionalized individual."[5]

The advocates of the new sciences of choice held that, even if humans were not fully rational, their choices could be. The age-old debate about the limits to reason had led to incomplete, irreconcilable views of human nature because it had asked the wrong question. Instead of asking whether people were rational creatures, the question should be, what is the best system for producing rational choices? The object of study needed to be the choice, not the chooser. Thus redefined, the problem of reason was ripe for a new synthesis, which, its proponents hoped, would provide both a rationale and a set of practical tools for the modern liberal state.[6]

These researchers reframed the old question of the limits to reason in a new language, a language of choices, decisions, revealed preferences, games, moves, payoff functions, subjective expected utilities, uncertainties, information flows, information costs, strategies, heuristics, programs, and the structures of cognitive processes.[7] As Edward Jones-Imhotep has shown, a closely related discourse about reliability in human-machine systems flourished in engineering at this same time, one that similarly reframed doubts about human capacities in a new language of systems design and error-measurement.[8]

Paul Erickson, Lorraine Daston, and their collaborators locate a similar shift in the discourse of reason in the Cold War era in *How Reason Almost Lost Its Mind*. They argue that a new "Cold War rationality" came into being in the years after Hiroshima, as the threat of nuclear war lent an "unprecedented sense of urgency" to debates about reason, rationality, and judgment. Thus, this new brand of rationality aspired to universality but was the product of a specific time and place—the United States in the Cold War, especially the "pre-detente" Cold War. It was "summoned into being in order to tame the terrors of decisions too consequential to be left to human reason alone."[9]

This Cold War rationality was different from earlier notions of reason and reasonableness, which (up through the Enlightenment, at least) assumed the exercise of judgment in context. Cold War rationality, in contrast, combined rule-based formalism, economic calculation of means to ends, experimental microcosms, and "towering ambitions." To an Enlightenment thinker, a machine—or a slave—could not exercise reason, for reason required both judgment and autonomy; to be rational meant to exercise good judgment, which required the freedom to make meaningful choices. Cold War rationality meant following rules and thus could be (and frequently was) embodied in machinery, either bureaucratic or electronic.[10]

The shift in the discourse of reason that Erickson et al. identify connects to the organizational revolution, as enacted during the Cold War. In brief, the development of the sciences of choice was shaped not only by local institutional contexts and Cold War fears (and patrons) but also by a new way of engaging with one of the central problems in modern social thought—the limits to reason and the consequences of those limits for democracy. This new way, the way of the high modern sciences of choice, was profoundly shaped by the context of an organizational revolution amplified and technologically accelerated by the Cold War.[11] This common context meant that, while there were many variations to these high modern sciences of choice, they shared a common theme: the hope that shifting the focus from the decider to the decision, from the person to the process, offered the possibility not just to save reason in theory but to produce it in practice. People might not be rational, but decisions could be. Rational decisions could be *made*.

The new sciences of choice promised to resolve the mid-century "crisis of reason" in a way that met positivist criteria for good science. They also sought to provide practical tools for managing large technosocial projects, which appealed to patrons with deep pockets.[12] After the mid-1970s, however, new challenges to the high modern sciences of choice arose, as ideas about choice shifted from being part of a discourse about reason to being part of a discourse about individual freedom.

The Sciences of Choice and the Crisis of Reason

Despite the increasing power of impersonal, "rationalizing" forces, such as the market and large-scale organization, social scientists from the turn of the twentieth century onward increasingly argued that human behavior was strongly influenced by nonrational beliefs and habits, most notably, by religion and local cultural traditions.[13] To the majority of social scientists of the 1880s–1890s, who had themselves been raised to value religion and its civilizing virtues, the importance of such subjective factors in determining behavior and belief was not necessarily a bad thing. The problem was less subjectivity than its decline: from where would values come in a secular world?

In the early twentieth century, however, the problem of subjectivity was reinterpreted. European social thinkers, such as Sigmund Freud and Vilfredo Pareto, ascribed the horrors of the Great War to the power of the irrational.[14] In America, the fear of unreason was less extreme, but it still led to myriad studies of the social disorganization that came when social groups clung to traditional values in the modern urban environment. The work of the Chicago School of Sociology is a perfect example of such concern: Robert Park was fascinated by the problem of "social control" in the ethnically diverse modern city, William I. Thomas studied the adaptive and maladaptive qualities of the culture of the *Polish Peasant in America*, and William Ogburn explored the problems of "cultural lag."[15]

The subjective aspects of human behavior similarly fascinated political analysts. The startling power of American pro-war propaganda and the manifest irrationality of public opinion attracted the interest of Charles Merriam and Harold Lasswell, for instance.[16] Works like Merriam's *Non-Voting* and Lasswell's *Psychopathology and Politics* revealed a corrosion of the iron faith of the older generation of political scientists: no longer could one assume that expert leadership and democratic politics could be reconciled through the education of a rational public. When the public was not apathetic it was only because its emotions had been manipulated.

Merriam and Lasswell were not the only political analysts dismayed by the irrationality of public opinion. Walter Lippmann's biting *Public Opinion* and his plea for expert leadership, *Drift and Mastery*, were widely hailed as taking a clear-eyed look at how democracy "really" worked.[17] From Watson's behaviorism to Freud's psychology to Pareto's sociology, theories that denied the importance (and even the existence) of the rational will excited the interest of political analysts. It was, as one historian of the period has termed it, the "crisis of democratic theory," a crisis rooted in a loss of faith in the rationality of the citizen.[18]

By the 1940s, a great many social scientists had become convinced that the ordinary human actor was (at best) imperfectly rational. Whether one was a Freudian or a Skinnerian psychologist, an opinion pollster or a market theorist, a culture-and-personality anthropologist or a functionalist sociologist, deliberate rationality no longer could be assumed of one's subjects. Even economists knew that *homo economicus* was not rational by any ultimate standard. As Jacob Marschak, one of the leading members of the famed Cowles Commission for Research on Economics, put it in one article, people just "do not behave rationally."[19]

At the same time, world events clearly showed the danger of unbridled unreason, and democratic ideals, which long had rested on faith in the mostly rational citizen, were not to be given up lightly. In the first two decades after World War II, the rising leaders of a new generation of social scientists sought a way to deal with this dilemma, hoping to find a safe ground for both their science and their society.

One result of this need to reassess the role of reason in human affairs was the explosion of interest in "decisionmaking." In field after field, researchers redefined their theory and practice around the study of decisions or choices. About the only other topic of quite such widespread interest at the time was the closely related subject of communication. System, that other great buzzword of the period, was less a topic of study than an assumption about the structure of the world: most scientists assumed that their objects of study were systems, whether political systems, economic systems, communication systems, decisionmaking systems, or information-processing systems.[20]

These three concepts—decisionmaking, communication, and system—often went together. As we have seen, to most postwar social scientists, systems were defined by their relations, not their elements, and communication and control processes were understood to be the fundamental relational mechanisms.[21] Many books and articles linked all three, studying the role of communications and information in decisionmaking, and vice versa, with the communications and the decisions taking place within a well-defined, interdependent system. Karl Deutsch's *The Nerves of Government*, for example, analyzed the role of a nation's communications patterns in shaping its citizens' decisions about their identities and about their governments' policies.[22] Going in the other direction, Claude Shannon's famed communications theory described a bit of information as that which enabled a receiver to move one step further down a decision tree in decoding a message.[23] Interpreting a message was a decision process; hence, language, communication, and even the basic interpretation of sensation all were decision processes, which was a new way to understand these very human activities.[24]

Similarly, Herbert Simon's organization theory was explicitly a theory of decisionmaking; he believed the primary reason why people organize themselves into firms, or any kind of formal organizations, was to enable them to make decisions that they could not make alone.[25] Earlier theories of the firm generally had seen it as means for accumulating capital or harnessing the power of the division of labor. To Simon, however, firms were means for accumulating the resources necessary for decisionmaking. And Simon was not alone: many proponents of the new "management science" or "administrative science," such as Peter Drucker, saw firms as decisionmaking entities first and producers of goods or services second.[26]

In the same vein, many political scientists in this period, such as Robert Dahl, Harold Lasswell, and David Easton, focused on political decisionmaking, from the choices of voters about their leaders (as in the epic studies of *The People's Choice* each presidential election) to the choices of political leaders about which positions to hold on which issues and which coalitions to join.[27] Even that longstanding preoccupation of political scientists, power, was redefined as that which influences decisions.[28]

Economics also embraced a decisionmaking framework, with utility theory redefined as the study of individual consumer choices and welfare economics as the study of how best to aggregate those choices in a group.[29] Choice long had been part of economic theory, of course, but the explicit study of choices as such, and the idea that the basic unit of economic activity was the *choice* rather than the *exchange* was new to the middle third of the twentieth century. At the same time, game theory, statistical decision theory, operations research, and systems analysis were defined from the start as sciences of choice, and they became more and more central to economic and political thought from the 1950s onward.

Studies of culture, especially of political culture, could be reframed as studies of how value systems influenced choices. Organization theory, learning theory, and social and cognitive psychology all could be reinterpreted in the same way.[30] As the organization theorist James March wrote in 1955, "students of a significant number of . . . types of behavior have tended to formulate their problems within a decision-making framework . . . [As a result,] there exist potential fruitful parallelisms among such theories as those of consumer behavior, administrative behavior, price setting, legislative enactments, propaganda, learning, foreign affairs, and social control."[31] Or, as Howard Raiffa and Duncan Luce put it in their widely used text, *Games and Decisions*, this theory of decisionmaking applied any time there was a conflict of interest to be resolved or a resource to be allocated.[32] Even the very process of scientific inquiry itself could be reframed in the language of choice: to those who studied Abraham Wald's statistical decision theory, science was a game against nature in which the scientist made experimental moves in order to decide on nature's true state.[33] In short, to many social scientists in the 1950s and 1960s, decisionmaking was a powerful new framework for understanding nearly all of human behavior.

Saving Reason

But how would the study of decisionmaking help social scientists deal with their dilemma? How could they maintain genuine hopes for democracy if their science continued to teach them that people were irrational? And, more immediately, how could they provide the necessary tools for managing complex organizations, if people were shaped by their irrational passions, not their rational self-interests?

The key was a subtle, but significant, shift in the basic unit of study. The new unit of study was the choice, not the chooser. The decisionmaker could be a person, a firm, a community, a legislature, a nation, even a machine—which meant that it really was none of these. If the decisionmaker could be a group as easily as a person, then those things peculiar to individual people—their unique life histories—could not be part of the analysis. Also, if the decision was the unit of analysis, then all decisions could be brought under the same umbrella: casting a vote, purchasing a tube of toothpaste, and going on strike all were choices subject to the same general analysis.

The chooser was a process, not a person, which meant that the analysis began in the present and looked forward, not back to the individual's past. For example, a Freudian psychotherapist would try to get patients to understand their past—to recover memories, traumas, repressed fears—in order to bring their psyches back into healthy balance. A psychologist trained in newer forms of cognitive therapy, by contrast, would try to get patients to think in terms of decisions and consequences; the past is not irrelevant, but it is more a source of data on which to base expectations of consequences than a living power shaping mind and behavior. Kenneth Arrow noted this difference in "Utilities, Attitudes, and Choices," stating that the "life history" approach is simply impracticable, while a narrower action-consequence framework is powerful and useful.[34]

Attention to this shift sheds light on many otherwise puzzling aspects of postwar social science. For example, it helps us see how social scientists could be so excited about the prospects of social science and social reform at the same time that their research seemed to show that people are emotional, illogical, poorly informed, and easily manipulated.[35] It also helps us understand how social scientists could be (at least seemingly) genuinely committed to liberal democracy even as they built institutions that seemed designed to constrain people's ability to make meaningful choices. Project Camelot, the infamous attempt to use social science as a tool to subvert revolutionary movements in the Third World in the 1960s, exemplified this fusion of cynicism and idealism, mixed with an unhealthy dose of political naïveté.[36] Similarly, the Vicos project in Peru, which produced influential studies of modernization and development, blended stunning confidence in the power of social science to bring about positive social reform with an equally startling belief that the lead researcher's status as the unelected, absolutist patrón of the community would enable the project team to study democracy in action.[37]

A more abstract parallel to such real-world projects can be seen in the "prisoner's dilemma" thought experiment. In the prisoner's dilemma, two prisoners are assumed to have been captured by the authorities and each has been offered a reduced sentence if he or she rats out the other. If both prisoners hold firm, then they go free, but if one talks and the other does not, the one who talks minimizes his penalty while the one who holds out winds up with the maximum sentence. The fascinating thing about the prisoner's dilemma is that many social scientists used it to understand the decision processes of people in all manner of choice situations, including the choices common to democratic self-governance. Yet the situation it models is an ideal-type authoritarian regime, in which the regime has reduced the range of choices to a simple binary—compliance or punishment—and in which the individual confronts the power of the state alone. Nevertheless, the advocates of the sciences of choice saw such analyses as crucial to saving reason and, therefore, democracy.

The shift from the chooser to the choice also helps explain why many articles in the economic literature of the 1940s and 1950s begin with a denunciation of outmoded, simplistic conceptions of the hyperrational homo economicus, only to follow this dismissal of the rational chooser with an analysis of rational choices. Indeed, a not unfair paraphrasing of dozens of articles from the period might run as follows: "Homo economicus does not exist because people do not behave rationally. Instead of assuming a rational individual, then, let us begin by assuming the existence of a set of ordered pairs, with the relations among these pairs being transitive; we then will test the implications of various postulates (corresponding to different decision rules) for this ordered set."[38]

What could be the point of combining such a dismissal of the rational chooser with such a study of rational choices? The point was to define what a rational choice was—to define reason itself—and to explore whether such a thing as a rational choice could be produced by *any* kind of system or process. If it could, then the next task would be to turn the description of the ideal into a norm for the real, creating systems, decision processes, policies, procedures, and protocols that would enable deciders to make rational decisions, whether those individuals were humans, machines, or groups of humans and machines.

Hence the power and persuasiveness of Arrow's *Social Choice and Individual Values*, in which he proved his "impossibility theorem" for social choice, one of the most inspiring negative results in the history of science.[39] This proof showed that no system for aggregating individual choices into a social preference function could meet certain basic standards for summing individual choices.[40] The implication, as understood in the 1950s and 1960s, was not that democratic politics was impossible but that individual choice inevitably must be constrained for a polity to exist: freedom does not exist without limits. As we saw in chapter 3, Arrow's student Anthony Downs applied a similar analysis directly to political issues in *An Economic Theory of Democracy*.[41]

Early rational choice theory required little conscious rationality of its subjects. It only demanded that individuals be able to decide whether they preferred A to B and B to C and (almost always) that these preferences be transitive.[42] This is simultaneously a remarkably low threshold for ratiocination (one does not have to "think" to prefer one thing to another) and an absurdly unrealistic expectation, implying that a limited chooser can evaluate infinite possibilities and their consequences.

The problems associated with this assumption attracted a great deal of attention and led to some clever solutions, such as Holt, Modigliani, Muth, and Simon's work at Carnegie Tech that (collectively) sought to show that the analyst could collapse all moves after the first into a single choice as to strategy, which dramatically reduced the "search space" for choices. Other efforts to scale back the demands for infinite information processing included Simon's idea that people "satisfice" but do not "optimize" (that they choose solutions that are good enough—usually the first one that satisfies certain criteria—rather than insisting on perfection), and his argument that the world is "nearly decomposable" (that it is "factorable" into nearly self-contained subsets, and hence all the implications for one's choice of factors outside the local subset usually can be ignored).[43]

But even these important emendations to choice theory rested on some debatable assumptions, probably the most important of which are that choices present themselves in universally identifiable, bounded units or "moves" (that everyone looking at a given situation sees the same alternatives A, B, C, etc., rather than certain people being able to see alternatives that others do not imagine), and that the act of choice is a selection among a set of such units (rather than the creation, combination, or redefinition of alternatives). These assumptions cause no great problem when the question at hand is a choice among brands of toothpaste, but they are troublesome when the question is one of grander strategy in an environment that is less "well defined."[44]

Despite such critiques, the study of decisionmaking offered its enthusiasts an opportunity to save reason while admitting unreason, a chance to validate their sciences, and a way to help reform society, all at once. It helped resolve the dilemma of reason because decisions could be rational, even if the decisionmaker was not. Social science could help people make better decisions by defining what a good, rational decision was and then by examining the processes that led to such decisions. It could compare this ideal decision process to the possibilities actually available to real people and use the difference between the ideal and the real either to guide people to a better decision path or to help social scientists achieve a better understanding of what rationality really was.

The instrumental nature of these goals is clear when one examines some of the key works of the period. *Decision Processes*, the edited volume that resulted from a joint conference for RAND, the Cowles Commission, the University of Michigan, and Office of Naval Research, for example, presents its abstract analyses as practical paths to rational decisionmaking.[45] Similarly, Herbert Simon and Allen Newell created a computer program that simulated human cognitive processes, treating cognition as a sequential process of selection from a set of operations.[46] Simon believed this simulation to be so powerful an abstraction that he sought to redefine the engineering curriculum at Carnegie-Mellon University around the teaching of what he called the "sciences of the artificial."[47] By *artificial*, Simon meant those things that might be other than they are—that is, things that could be chosen. To him, the sciences of the artificial were simply the flip side of the sciences of choice.

Producing Reason

The end goal of the technosocial sciences of choice was the design of systems that would generate rational choices automatically, whether the humans involved were rational creatures or not. Rational decisions would be *produced*.

The goal was shared by new societies such as the Operations Research Society of America and the Institute of Management Sciences, new journals such as *Administrative Science Quarterly* and *Management Science*, new schools of business and of government, and a new species of business strategy consultant such as McKinsey & Company. To those who doubted that such procedures could work, the proponents of the sciences of choice replied: workers on the assembly line do not have to be tremendously skilled or highly rational to turn out precise, standardized products by the million. The system harnesses the capacities they *do* have so that they can, as part of the system, accomplish something they could not do individually. People do not have to be rational for work to be rationalized. Decisions could be produced, just like any other kind of widget; there were costs associated with their production (costs of acquiring information, in particular), just like any other production process.[48] White-collar work could be automated to reduce such costs via the "new science of management decision" in much the same way that physical costs had been by the assembly line.[49]

Although the sciences of choice, especially game theory and operations research, could be fantastically abstract, this was an instrumental approach to social science. The desired product almost always was what Daniel Bell called an "intellectual technology."[50] Such intellectual technologies could be algorithms for solving certain classes of problems; they could be policies and organizational procedures, flow charts, decision rules, protocols, computer programs, and more. Tellingly, Herbert Simon and Allen Newell called all such abstract yet instrumental systems of "if . . . then" rules "production systems," whether the system took biological, mechanical, mental, symbolic, organizational, or social form.[51]

From these attempts to create production systems for rational choices came many of the tools and concepts that we have used to organize our world over the past sixty years, from the algorithms for solving the "traveling salesman" problem, which helped give us the hub-and-spoke arrangement of our air travel networks; to the queueing theory of Leonard Kleinrock and the communications protocols (TCP/IP) of Vinton Cerf and Robert Kahn, which paved the way for the Internet; to the linear programming and statistical quality control techniques that enable "just in time" auto manufacturing; to the rules for optimal inventory control that keep our grocery stores' shelves stocked; to the entire armada of policy evaluation techniques (especially cost-benefit analysis); to psychological counseling techniques that focus on improving the client's ability to "make good choices."

The transformation of schools of business and of government was a vital part of this reconstruction of society through the sciences of choice. Beginning in the 1960s, graduate training in business, management, and administration and government was reoriented around the tools, techniques, and concepts of the sciences of choice. Thanks to the Ford Foundation's $35 million program to reform business education, Carnegie Tech's Graduate School of Industrial Administration (GSIA) became the model both intellectually and institutionally for new MBA or MA in government or administration, a model embraced by many leading programs, such as the ones at Stanford, the University of Chicago, and the University of Texas, among others. By the 1990s, such schools of administration trained 100,000 students a year in the sciences of choice and the techniques of administrative control.[52]

There was—and still is—an enormous variety of approaches to the study of decisions and choices, and the differences between these approaches could be large. In particular, experimentally oriented behavioral scientists, especially cognitive psychologists like Herbert Simon, Jerome Bruner, George Miller, George Katona, Amos Tversky, and Daniel Kahneman, diverged sharply from the exponents of abstract rational choice theory, especially game theorists and structuralist mathematical economists such as Kenneth Arrow. Cognitive psychologists, by and large, studied decision processes in groups, individuals, or machines with the middle-range goal of modeling how humans actually make decisions. Game theorists and structuralist mathematical economists, by contrast, constructed abstract models of choice processes and systems, testing the implications for these models of various limiting conditions, rules for combining preference sets, and criteria for rationality.

Despite these very real differences, the varied sciences of choice did share certain features that were new to postwar social science: First, and foremost, they began with the idea that the choice or the decision was the thing to study. Second, they agreed that decisions were the products of sequential (sometimes iterative) processes, with each step in the process being a selection from a set of alternatives; (hence the fascination with "courses of action," "decision processes," flow charts, strategies in games, Markov chains, algorithms, heuristics, and, of course, computer programs). Third, they shared an expectation that such processes could be modeled formally—indeed, they all tended to believe that constructing models is what scientists do—and that these formal models eventually would aid in the rationalization of human choices. In short, they were as optimistic about the power of organized reason as they were pessimistic about the overall rationality of the individual human.

Reason and Democracy

There are many ways to address the question of the role of reason in human affairs, many ways to reconcile human flaws with democratic governance. A defense of democracy need not rest on the rationality of ordinary people; indeed, if rationality is defined in terms of the efficient pursuit of narrow self-interest, as it so often is, then democracy may well depend on people being irrational and sacrificing something they value for the good of the whole.[53]

Why did the postwar generation address this question the way they did, then? Why did they shift from the decisionmaker to the decision, why did they develop such extensive production systems (mental and organizational) for rational choices? Why was there such intense, widespread interest in creating a science of choice, and why did that science take the forms it did?

This explosion of interest in decisionmaking was the product of the confluence of at least three factors. First, approaching the question of the role of reason in human affairs via the analysis of decision processes fit well with the bureaucratic worldview of high modern social scientists and their patrons.[54] The basic assumption at the heart of this worldview was that the world is a complex, hierarchic system. Science was about modeling the processes and relations that define such systems. The exponents of this view embraced what one might call a "broad church" positivism; that is, they believed that the data of science must be observable and its concepts operational, but they held a variety of views about how to deal with unobserved "intervening variables."

The result was an intense enthusiasm for mathematical, behavioral-functional social science, especially for formal mathematical models of the processes and relations that defined systems. The fascination with decision-making fit well with this approach because, while a person's mind cannot be observed, his or her decisions can be. A classic example of such reasoning is Paul Samuelson's famous paper on "revealed preferences."[55] In it, Samuelson sets aside questions of what people really want (their individual absolute preference schedules) and focuses on preferences as revealed through concrete choices. Studies of concrete decisions, in turn, could give insight into the processes that generated them and the nature of the decisionmaking system.[56] Similarly, one of the reasons for the fascination with both small group study and with computer modeling was that they promised to open up the decision process for observation, enabling the experimenter to see acted out the steps in a decision process that normally went on inside the subject's head. Individual psychology was thus, as Allen Newell once put it, "organization theory in miniature" because both were, at their core, theories of decisionmaking.[57]

Second, in the early postwar period many social scientists, natural scientists, business leaders, and military leaders believed that the world had grown dramatically more complex and decisionmaking exponentially more difficult. The scale of organizations, both public and private, had increased so much, and the rate of production of new knowledge had grown so rapidly, that old methods for making decisions or dealing with information seemed obsolete. As Vannevar Bush wrote in 1945, man "has built a civilization so complex that he needs to mechanize his records more fully if he is to push his experiment to its logical conclusion."[58]

At the same time, the advent of extraordinary new technologies, such as nuclear weapons, meant that this was a bad time for bad decisions. New techniques (and new technologies) were needed to help us humans make good decisions despite our limited capacities. Improving communications was part of the solution to this problem, since communication lines defined the "problem-solving" or decisionmaking organism. Improved information processing technologies also were part of the solution, and so was the study of decisionmaking.

Third, there were many new patrons interested in solving these questions of communication, choice, and control in large organizations. For economics and game theory, the most important early postwar patrons were the ONR and

RAND. They funded an enormous amount of work on decisionmaking directly and through the Cowles Commission, whose projects on "decision-making under uncertainty" (led by Jacob Marschak) and the "theory of resource allocation" (which was all about rational allocative choices) supported multiple future Nobel Prize winners, including Kenneth Arrow, Gerard Debreu, Tjalling Koopmans, Franco Modigliani, and Herbert Simon.[59] In addition, the ONR supported the Behavioral Models Project at Columbia University's Bureau of Applied Social Research, headed by Paul Lazarsfeld. The Behavioral Models Project staff included Duncan Luce and Howard Raiffa, authors of the crucial exposition of game theory, *Games and Decisions*.[60] If one searches an online database such as JSTOR for articles from the 1950s having to do with decision-making, one is over 90 percent likely to find the author of the piece was at least partially sponsored by the ONR or RAND.

Other foundations and research agencies also supported work on decision-making and the sciences of choice in the 1950s, including the Rockefeller Foundation, which supported the Social Science Research Council, which, in turn, had a Committee on Research on the Business Enterprise that was enthusiastic about decision science. Probably the most important civilian source of support in the late 1950s and early 1960s was the Ford Foundation, which funded the work of the researchers at Carnegie Tech's GSIA (including Herbert Simon, Franco Modigliani, Charles Holt, James March, Allen Newell, and G. L. Bach). Also, as noted earlier, the Ford Foundation's $35 million program to reform business education, which had a major impact upon the way that MBA programs were restructured in the 1960s, took the GSIA as its model and "management science" (the sciences of choice) as its intellectual core.[61] All of these patrons saw interdisciplinary, instrumental, technosocial science as vital to solving the new problems of managing a complex world.

Conclusion

The irony of this story is readily apparent; by shifting focus from the decider to the decision, one vested power in the system that produced the choice rather than in the individual doing the choosing. Hence, while many of the proponents of such decision systems genuinely believed that they were helping to shore up democracy, their critics saw them as restricting freedom. By the 1970s, many people began to feel trapped and voiceless within such technosocial systems and to resent the narrow "rationality" these systems produced.

An abstract, but telling, example of this problem is the Prisoner's Dilemma. As noted above, this thought experiment is, in theory, a universal model of rational choice, but it is a model of choice in an authoritarian situation. The authoritarian element is most prominent in the limited conceptions of time, communication, and power one finds in the standard versions of the Prisoner's Dilemma: most models are of decisions in situations where the participants lack the ability to communicate with each other or to imagine a future beyond the moment. People in such situations have no prospect of future relationships with each other. Real life, of course, is radically different. Given the ability to communicate and the prospect of future interaction, one can make quite persuasively the age-old argument that morality in the sense of self-sacrifice for the larger, longer-term good is extremely rational.

In addition, political power plays a strange role in these calculations. Why is teaming up with the other prisoner to kill the guards and take over the prison never an option? Why are they prisoners? Why does the core model of this kind of political science assume that its subjects lack the power to rebel, which is a fundamental aspect of real political calculations? A regime that imposed the conditions of the prisoner's dilemma on its citizens would be a totalitarian system beyond Stalin's dreams: a situation in which every citizen is forced to act only as an individual with respect to the awesome power of the state and in which he or she is faced with a continuing series of "betray your neighbor or die" choices. When you cannot win, you cannot break even, and you cannot get out of the game—well, that does not sound like freedom.

Such a constrained concept of reason—abstract in the Prisoner's Dilemma but all too tangible in the myriad procedures and protocols of the applied sciences of choice—is one reason for the broader shift in ideas about choice shifting from being part of a discourse about reason in human affairs to being part of a discourse about individual freedom.[62] The speed and scope of this shift is startling—as is our continued dependence on the sciences of choice for maintaining the large-scale organizations that have become vital parts of our technosocial infrastructure.

This shift from reason to freedom as the context for understanding choices also has led to a valorization of the act of choice itself. Certainly we are taught to make good choices, but it is striking how often people today are lauded for the intentionality rather than the rationality of their actions. It is also striking how rationality has "receded": for conservative defenders of free markets, the market is the source of reason, while for more liberal defenders of social free-doms, choice often is seen as part of breaking free of the straitjacket of reason, rather than as a product of a broader rationality.

In addition, while the sciences of choice did offer new ways of understanding creativity and emotion, these often were rather cramped visions, as one can see through the work of Jamie Cohen-Cole on creativity and of Marga Vicedo and other scholars of the history of emotion.[63] To the sciences of choice, creativity involved the generation of new alternatives from which the chooser could select (making it a kind of meta-selection process), and emotions were either givens (like values or preferences) to be input into decision processes (as specifications of parameters, say) or intrusions that short-circuited the normal processes of decision (for good or, more commonly, for ill).[64] This was an ironic result indeed, for the exponents of the postwar sciences of choice were both creative and passionate in their attempts to produce reason and, as they saw it, save democracy.

CHAPTER FIVE

Modernity and Social Change in American Social Science

We are experiencing one of the great revolutionary transformations of mankind . . . the resulting patterns of change offer unprecedented prospects for the betterment of the human condition, but at the same time threaten mankind with possibilities of destruction never before imagined.

Cyril Black, 1966

If there is any constant in American social science over the past century, it is the awareness of change, its possibilities, and its discontents captured in the opening paragraph of Cyril Black's widely read *The Dynamics of Modernization*. Indeed, as Dorothy Ross put it so clearly in *The Origins of American Social Science*, "The discovery of modernity is the fundamental context of the social sciences."[1]

This discovery of modernity was not always a happy discovery. It was not just social critics, such as Karl Marx, who saw everything solid melting into air or artists, such as Gertrude Stein, who saw everything in the twentieth century "destroying itself."[2] It also was sober and conservative social observers, such as William Ogburn, who stated, "Modern society has not made a successful shift to a condition of change."[3] Nevertheless, whether they liked or disliked the change they saw all around them, American social scientists in the twentieth century understood their world to be a new one: a world changed, fundamentally from the "world we have lost."[4]

But did they all discover the same modernity? Were the crucial features of the modern world and the central problems the same for all (or even most) social scientists from Ogburn's day to the present? Or was there an evolution in thinking about change? A revolution? And if there were changes in thinking about social change, in what ways were these new ideas the products of the very changes they were created to analyze?

Modernization theory neither arrived nor departed unnoticed. From the beginning, it spurred controversy and criticism, though those critiques began at the margins of power and belief and only moved to the mainstream during the Vietnam era. It has continued to receive attention from historians today, whether they be scholars of the history of empire, such as Michael Adas; of the Cold War, such as Michael Latham, David Engerman, or Nils Gilman; or analysts of the postmodern, global condition, such as Daniel Bell, David Harvey, or Manuel Castells.

The most relevant histories, for the present analysis, are those that situate modernization theory squarely within the context of Cold War America: Latham's *Modernization as Ideology*, Engerman's *Modernization from the Other Shore* and *Know Your Enemy*, Gilman's *Mandarins of the Future*, and the collected volume (co-edited by all three) *Staging Growth*. These works all are grounded in the idea that the most relevant context for understanding modernization theory is American social science during the Cold War. All three scholars acknowledge modernization theory's roots in older ways of thinking and lament its lingering power today, years after it was declared dead by its critics, but their accounts of modernization theory's rise and fall depend on the unique circumstances of Cold War America.[5]

This strong link of ideas to specific context is by no means a flaw; it is, rather, the essence of good intellectual history. All of these excellent works are products of deep knowledge of modernization theory. They present hard-won conclusions about modernization theory being an ideology peculiar to Cold War America, connected to liberal change abroad and conservative consensus at home, and dependent on a largely unexamined faith in the future convergence of modernisms on an American-style high-technology, high-consumption, social welfare state.

However, viewing modernization theory through a slightly different, wider-angle lens—putting it in the context of social scientific thought about modernity and social change over the course of the twentieth century—illuminates not only the continuities in thinking about social change but also those aspects of modernization theory that were indigenous to the high modern social science of the first decades after World War II. And it continues to apply in more current theories of globalization and development, as in the work of Daniel Bell, Manuel Castells, and David Harvey.

High modern ideas about modernity demonstrated both continuity and change: continuity from the prewar through the early Cold War periods in the basic distinctions between traditional and modern societies and in the grand narrative of development; change in how the process of change was understood, in the institutions and methods through which it was studied, and the means by which research was linked to practice. Later analyses of modernization have retained certain key elements of the modernist tradition but have been much more conflicted about the grand narrative of development, partly because they see a loose coupling between the economic, political, and cultural aspects of the modern world, and partly because there is no longer a consensus that modernization on the American model is universally desirable.

Continuities

If one wants to understand how the Cold War context and high modern style helped shape concepts of modernity and social change, the first thing to realize is that pre–World War II and early Cold War ideas about modernity and social change shared a great deal. The distinctions social scientists of the era drew between traditional and modern societies and the grand narratives of the rise of Western modernity they articulated were much the same, both in their broad outlines and in many important particulars.[6]

Specifically, they viewed modern societies as controlling nature rather than being controlled by it, as evidenced by their superior level of material culture, especially their production, transportation, communication, and military technologies. This is a very old distinction, as Michael Adas has shown in his *Machines as the Measure of Men*, one dating back to the late Enlightenment, when Europeans first began to think of themselves as superior to the rest of the world not just because they were Christian but because of their Newtonian worldview and rapidly advancing mechanical technologies.[7] Indeed, many twentieth-century writers on modernity echo Alexander von Humboldt's sympathetic, but strongly Eurocentric, view of native peoples as being defined by their geography and climate, with only European peoples having broken the limits of nature on culture.[8]

They also saw modern societies as experiencing constant change and disequilibrium rather than undergoing only slow, spasmodic changes amid general harmony, a condition they saw as characteristic of traditional societies. William Ogburn, for example, contrasted "stationary" and "changing" societies (though he noted that, thanks to the intrusion of modern societies on traditional ones, there are no stationary societies left in the world). Similarly, Cyril Black wrote of modernity as fundamentally a "dynamic" condition, one that needed a properly dynamic theory to be understood. Even Walt Rostow, who took pains to note that "the story of traditional societies is a story of endless change," immediately followed up that point by setting a rather low "ceiling" on that change.[9]

This emphasis on modernity as a state of constant change entailed the development of concepts like "cultural lag," and it led to an acute awareness of the modern need for continuous social and psychological adjustment, raising the perpetual question of what will bring order, meaning, and purpose to dynamic societies and their members?

In addition, both prewar and high modern, Cold War writers on modernity saw it as being characterized by the penetration of instrumental reason throughout almost every aspect of life, a change in outlook that stood in sharp contrast to the rule of religion, custom, and arbitrary personal authority they saw in traditional societies. In this, American writers followed a long and famous European tradition, one that included Marx's critique of capitalism as reducing all social relations to the cold calculation of commodity exchange, Ferdinand Tönnies's distinction between the *Gemeinschaft* of personal, historically grounded relationships and the *Gesellschaft* of impersonal, rationalized connections, and Max Weber's ambivalent analysis of instrumental reason's rise from its Protestant cradle only to imprison its mother in the "iron cage" of bureaucracy.

For these authors and their American followers, the spread of instrumental reason dissolved old cultural norms. This dissolution encompassed the secularization of society; the breaking of the bonds of custom, including the customs of rank, deference, and racial and sexual hierarchy; and the severing of the emotional and psychological ties that bound individuals together into communities. Hence the rise of individual freedoms constantly was paired with increases in psychological dislocation and alienation: modernity was not for the weak. Only the tough-minded (in William James's phrase) could find purpose and meaning in the world of the "homeless mind."[10] Indeed, for people often described as triumphalists, the social scientists who wrote about modernity and social change wrote a great deal about dislocation, disintegration, alienation, and violence as products of modernization. In the end, they thought modernity was worth the price (or, at least, that it was less expensive than the alternatives), but their view of modernity only looks sunny when compared to the soul-wracking anguish of modernist artists.

A final aspect of the spread of instrumental reason was the rise of (Western) science and technology. High modern writers, both prewar and postwar, agreed that here was the great moral hope of modernity, that material ease might loose the bonds of self-interest, cut the "cake of custom" and tribalism, and allow reason to inform moral judgment. Thus, while instrumental reason would bring disorder by dissolving the ties that bound traditional society, it also would bring the promise of science as the solution to the problems reason had caused.

They also agreed that modernization was a total change in a society, not just a change in economic relations or technological gadgetry, though economic and technological changes were the easiest to observe. Indeed, as Ogburn's article "Culture and Sociology" reminds us, one large part of the appeal of the term *culture*, both before and after World War II, was as a way of capturing the wholeness of that set of interrelated features that distinguish a traditional society from a modern one.[11]

Some authors, including Ogburn and Rostow, saw economic and technological changes as the engines driving this total change in society, but even they argued that modernization was not a purely economic or technical process. (Black places the engine of change even farther back, in the cultural shift toward instrumental reason that followed the Newtonian revolution.) Ogburn, for example, coined the term *cultural lag* as a way of getting at the interconnectedness of the various aspects of social change, and Rostow stated quite directly that "the sectors of society interact: cultural, social, and political forces, reflecting the different facets of human aspiration, have their own authentic impact on the evolution of societies, including their economic evolution. They are not a superstructure derived from the economy."[12] This statement is the centerpiece of Rostow's argument against Marxian interpretations of modernization, which, he believes, treat all "human behaviour as an exercise in profit maximization."[13]

Black likewise describes modernization as a total process, noting that while "political scientists frequently limit the term 'modernization' to the political and social changes accompanying industrialization," a "holistic definition" is better suited to the "complexity and inter-relatedness of all aspects of the process."[14] Thus, to Black, modernization is even bigger than the vast set of social and political changes that accompanied industrialization—a total process indeed!

Finally, these analysts of modernity agreed on the broad outlines of a grand narrative of the rise of the post-Enlightenment West and the spread of Western culture to the rest of the world. In this view, all societies were traditional once, despite their surface variety. Due to various historical accidents, however, Europe, led by England, began to modernize in the late eighteenth century, eventually forcing the rest of the world to follow the path it had blazed. From the 1930s on, American social scientists consistently described the rise of the West as contingent even though modernization was inevitable; Europe, in this view, had certain cultural traits that enabled its peoples to take advantage of certain historical accidents of geography (cheap coal and easy water transport) and history. This view stood in sharp contrast to earlier American social science, which had a powerful racist and eugenicist strain and often cast the industrial revolution and American democracy as the triumph of the "Teutonic peoples."[15]

In this grand history, modernization was, and is, inescapable, once started, though many have tried to evade its grasp and avoid the violence and social disorder it inevitably brings. From this history, the logical conclusion that both prewar and early Cold War writers drew was that the task of the social scientist is to help a society adjust to the inevitable coming of modernity with as little pain as possible. How that was to be done remained a topic of disagreement, with social scientists taking a range of positions on a spectrum between seeing themselves as educators of the citizenry and seeing themselves as advisers to a governing elite.[16]

Changes

To agree on the all the above is to agree on a great deal; these were not empty generalities. But there were several significant differences between the ways that prewar and high modern social scientists depicted the world, even given this broad common frame and shared basic palette of concepts and distinctions. Four of these differences are very significant: First, prewar analysis of modernity and social change almost always was an analysis of social evolution, while during the Cold War it almost always was a study of social development. Evolution and development are closely related ideas that are frequently blended, but they are not synonymous, and the differences between the two indicate distinct modes of thinking about change.

Modernization theorists, for example, frequently used the word *evolution*, but they almost always used it to mean simply "continuous change over time" or a "continuous process of adaptation of the social organism to changed conditions." Neither of these ways of understanding evolution necessarily conflicts with a more thoroughgoing evolutionism, but both are more typical of developmental thinking than of twentieth-century evolutionary thinking.

Modernization theory was a developmental theory, a theory of the development of an organism according to an ingrained plan. It was a process that led through necessary stages of growth, from the childhood of traditional society to the painful adolescence of the "takeoff," to the hard work and sacrifice of early adulthood's "drive to maturity," to the masterful, integrated adulthood of full industrial maturity. (There is even a whiff of senescence in Rostow's "age of high mass consumption.")[17]

This developmentalism was most prominent in what Nils Gilman has identified as perhaps the strongest of all the unexamined assumptions of the modernization theorists: the belief that modernization was a convergent process. Modernization viewed as a developmental process easily could be seen as converging on a universal adult state; if it were viewed as a Darwinian evolutionary process, however, that convergence would demand explanation—and probably defy it. Evolutionary thinking, after Darwin at least, was about explaining patterns of differentiation and divergence every bit as much as it was about explaining adaptation to circumstance.[18] Convergence on forms particularly well suited to common environments is conceivable, but applied to the social world, such convergence simply begs the question as to how the environments converged.

Thus, while evolutionism and developmentalism went together in the nineteenth century, the harder edge given to evolutionary thinking after August Weissmann led social evolutionists such as Ogburn to scoff at developmentalist "absurdities" such as "the successive stages theory."[19]

It is true that modernization theorists wrote at length about the highly differentiated, specialized nature of social and economic functions in a modern society, a topic one might think was a natural venue for evolutionary analysis. They did so, however, in the manner of a physiologist remarking on the fine organization of specialized functions in a complex organism.[20] Such analyses are similar to, but different from, the way an evolutionary biologist would look at the mechanisms by which divergent forms were created over time. Tellingly, even that quintessential product of evolutionary population studies, demography, was translated into developmental language by modernization theorists: the demographic transition, to them, simply described the life cycle of a social organism as it matured from childhood (high birth and death rates) to adolescence (high birth and low death rates) to mature adulthood (low birth and death rates.)[21]

Prewar writers, by contrast, were part of an evolutionary discourse, whether they were Darwinians, eugenicists, Lamarckians, or simply sociologists. Thorstein Veblen wrote a famous article at the turn of the century, "Is Economics an Evolutionary Science?"[22] To him, that question was tantamount to asking if it was a real science. Much the same held true through the interwar period. Social evolution, and its relation to biological evolution, was one of the central questions in social science, as seen in Ogburn's famous text, *Social Change*, and the processes that generated the extraordinary division of labor in modern society were a core subject of study.

The irony of this shift from social evolution to social development is that modernization theorists argued fiercely for the need for a "dynamic" social theory to explain modernization, one that they saw themselves as providing. Rostow, for example, devoted the majority of his long essay review of *Toward a General Theory of Action*, by Parsons, Shils, et al., to the problems inherent in what he saw as its static model of society.[23] Developmental thinking could produce a dynamic theory, but it placed bounds on change that evolutionary thinking did not.

A second major difference was that Cold War–era modernization theory placed a far greater emphasis on the state as an agent of change than prewar studies of social change did. Ogburn, for example, does not mention the state at all in his best-known articles on social change, and it scarcely appears in *Social Change*. Only with the advent of the New Deal did Ogburn take note of the power of an activist state, but even then he saw that state as acting to aid society in adjusting to already changed conditions—that is, as helping solve the problems created by cultural lag, not as creating change in the first place.[24]

In modernization theory, by contrast, the state modernizes: it is the primary unit of analysis and one of the key engines of change. Whereas prewar thinkers highlighted the role of religion in fostering, opposing, or tempering the modernizing process, modernization theorists tended to ascribe that role to the state and political leadership. Religion thus became one more aspect of the traditional culture that modernizing elites had to change, often through the use of state power.

This emphasis on the state as unit and agent was a novel concept to postwar social science, and it fit rather uneasily next to traditional American commitments to individualism and private enterprise. It did, however, integrate quite well with modernization theory's developmentalism, for the state could be cast in the role of that which gives a direction to the developmental process. It also matched recent historical experience, especially the Soviet Union's forced march to modernity and the success of the American welfare state in dealing with the crises of economic depression and war. And, perhaps most importantly, it met the needs (and self-image) of policymakers in the newly active, interventionist state, a new class that liked theories that ascribed historically crucial roles to policymakers just like themselves.

Third, modernization theory also distanced, universalized, and valorized the modernizing process by making stronger connections between today's modernity and the scientific revolution of the seventeenth century, as opposed to more narrowly European cultural events, such as the Reformation or the French Revolution. In this view, the scientific revolution was understood as an epic transformation of the world, the beginning of all modernity. While it originated in Europe, it was now global common property in a way that Protestantism or parliamentary democracy might never be. It was Europe's one great gift to the world it had colonized. Black, for example, defined modernization in relation to the scientific revolution, stating that it refers to "the process of rapid change in human affairs since the scientific revolution," a process that centers on the adaptation of "historically evolved institutions" to "the unprecedented increase in man's knowledge, permitting control over his environment, that accompanied the scientific revolution."[25] Black put this revolutionary transformation on par with the emergence of the first hominids hundreds of thousands of years ago and the development of settled agriculture in the Neolithic revolution, echoing Herbert Butterfield's famous claim in *The Origins of Modern Science* that the scientific revolution reduced such epic events as the Reformation and the rise of capitalism to mere side stories.[26]

Rostow was less grandiloquent, but his argument was much the same: "But limitations of technology decreed a ceiling beyond which [traditional societies] could not penetrate. They did not lack inventiveness and innovations, some of high productivity. But they did lack a systematic understanding of their physical environment capable of making invention a more or less regular current flow, rather than a stock of ad hoc achievements inherited from the past. They lacked, in short, the tools and outlook towards the physical world of the post-Newtonian era."[27]

Ogburn and his generation, by contrast, did not discuss the scientific revolution as part of their analyses of social change, except as a distant backdrop.

For example, in the 1930s, Ogburn's fellow sociologist Robert Merton famously took the scientific revolution as the subject to be explained through historical sociology, not the agent that explained all other social transformations.[28]

Fourth, in keeping with its close institutional, intellectual, and political connections to the state, early Cold War modernization theory had a much stronger operational component than did prewar social science, a fact that shaped both its intellectual and political history. To give but one example, the best known exponent of modernization theory (Rostow) was President Lyndon B. Johnson's National Security Adviser, which mattered both for the prosecution of the Vietnam War and for the rise and fall of modernization theory's appeal.[29]

This operational component was not completely new: social scientists had been involved in domestic policymaking during the New Deal (though only to a limited extent), and they had played such an important role in World War II that Paul Samuelson boldly proclaimed it to have been the "the economist's war" just as much as it was the physicist's.[30] Despite these beginnings, the large-scale, institutionalized integration of social scientific expertise into governance came after the war, with the creation of a permanent Council of Economic Advisers (with actual offices in Washington), the ascent of the Bureau of the Budget (later, the Office of Management and Budget) and the Fed, the development of large-scale social surveys and opinion polls, and the staffing of both line agencies and congressional committees with policy analysts with academic credentials.[31] Rostow's position as national security adviser was extraordinary but not anomalous.

This penetration of academically trained experts into the world of policymaking has continued to the present day. As the world of public policy experts has grown, however, it has become more distinct from the world of professional academics. There are some well-traveled bridges between these two worlds, and many public policy experts still receive their advanced degrees from traditional academic programs, but the worlds have grown apart. Public policy experts, for example, increasingly have received their advanced degrees from professional schools, such as schools of government or business or law, many of which have become centers for applied social research. In addition, the public policy expert's world has become professionalized, with its own set of institutions parallel to those of traditional academia: there are journals and centers devoted to policy analysis and applied social science, not to mention an ever-growing number of think-tanks. Due to its closer connection to specific policies, the tension between advocacy and objectivity has been strong in this new domain.

The growing distinction between the public policy expert and the professional academic began to matter for analyses of modernization and development around the end of the Vietnam War. Since that time, public policy experts, generally speaking, have retained their faith in the American model of development (though they no longer believe that it is the only viable one economically). Professional academics in this period, by contrast, have expressed a variety of opinions on American-style development, with many of the most prominent voices making sharp criticisms of modernity, American style. By the late 1980s, these differences had grown so marked that the collapse of the Soviet Union was understood in very different ways by members of these two groups: the typical public policy expert celebrated the fall of the Berlin Wall as confirmation that the American model had been right all along, while the typical academic analyst of development greeted the end of Soviet communism with at most two cheers—there were too many other, bigger problems facing us for exuberance to be rational.[32]

Contexts

The new road linking social scientists to government in the early Cold War ran the other way as well, with state patronage playing a powerful new role in social science. As shown in chapter 2, shifts in patronage for the postwar behavioral and social sciences were tied to both intellectual and institutional changes. There were two distinct, successive patronage systems for postwar social science, not one, as is commonly assumed. The first system played a major role in enabling a series of behavioral revolutions and interdisciplinary syntheses across the social sciences while the second encouraged the development of specialized concepts, techniques, and technologies *within* the disciplines.

The rise and fall of modernization theory fits this model almost perfectly: it was a consciously interdisciplinary enterprise based in the research centers created by the first patronage regime. It rose and fell precisely when other such centers did: growing rapidly from the end of the war through 1958, even more explosively from 1958 through the late 1960s, and then stagnating between 1968 and 1975, often with the result being dissolution of the center in question unless work could be supported on a new financial basis.

Modernization theory's career also entwines with the intellectual movements associated with these institutional changes: it was based on the behavioral-functional analysis of social systems/structures, and its practitioners sought to create (relatively) formal models of patterns of relations, seeing such social relations as the core objects of study. To them, the relation was the thing and the model the goal, which is not the only way to think about social science.

As a leading exemplar of such behavioral-functional analysis of social systems, modernization theory was party to both the rapid rise and savage critique that all such approaches experienced. Like other such theories, it emphasized the need for integration, order, and stability within social systems (which became unwelcome signs of political conservatism in the 1970s), and, despite its developmentalism, it had the same difficulty in dealing with change in the fundamental nature and structure of social systems that other systems theories generally suffered. Modernization theory, like most systems theories, could be quite helpful when it came to interpreting comparative statics—comparing the structures and functions of two systems or of one system at two different points in its history—but it was inadequate for explaining the dynamics of how one system transformed into another. Hence the consistent fuzziness of modernization theorists as to how, exactly, modernity emerged from the violent, chaotic period of transition that followed traditional society's collapse.

Like others involved in the interdisciplinary, behavioral-functional analysis of social systems, modernization theorists believed in both the need for objective social theory and the need to apply theory to the solution of concrete social problems. They held an experimentalist's view of the progress of knowledge and of the improvement of the world, one that saw greater knowledge of the world emerging from conscious, systematic intervention in it. Hence, the application of social theory to the problems of modernization—and even the enshrinement of a theory as official policy and installation of theorists as policymakers—need not compromise the objectivity of the theorist or the validity of the theory. Rather, that application actually could help researchers test their theories against experience and so make them more objective. There might be dangers involved in such an applied enterprise, of course, but modernization theorists, like most of their generation of social scientists, were startlingly unconcerned about such dangers until the controversy over Project Camelot erupted and the succession of horrors in Vietnam began to unfold. Members of the older generation were more leery of involvement either in policymaking or in federally funded research, as is evidenced by Ogburn's tepid testimony in support of the inclusion of the social sciences in the National Science Foundation's mandate.[33]

All of these intellectual changes occurred in an even more radically changed geopolitical context. For modernization theorists, there were two novel aspects of Cold War geopolitics that were vitally relevant to their mission, and to their understanding of the world: the first was the explosion of new nations casting off imperial rule. The second was, of course, the global struggle between superpowers. While this superpower struggle had much in common with earlier contests among great powers, it was characterized not only by competing strategic interests (common to all great power struggles) but also by two competing ideologies. Both of these ideologies were deeply modernist, but they reflected radically different aspects of modernism and institutionalized different cultural, social, economic, and political norms for engaging with the challenges of modernity. Hence, modernization theory, as Michael Latham points out, was much more of a consciously anti-Marxian ideology than were prewar theories of social change (which were always political, though not necessarily ideological).[34]

Conclusion

To what extent was modernization theory a product of the Cold War, and to what extent was it old wine in a new bottle?

The story is one of a long tradition of thinking about social change, organized around some basic assumptions and common experiences, within which existed great variety, both conceptually and practically, that became formalized and institutionalized in the new context of the Cold War. This new context produced, at first, a seeming consensus on modernization and modernity. While most of the principles underlying this seeming consensus fit with the broad intellectual tradition of American thinking about social change, the belief in convergent development, changing assumptions about the role of the state and political leadership in modernization, the embrace of a high modern bureaucratic worldview, and the idea that engagement with policymaking would improve the objectivity of science, as well as the efficacy of policy, all ran counter to important elements of that tradition. Hence, modernization theory encountered criticism from the beginning, was never monolithic, and was difficult to sustain, while the preexisting assumptions about modernity and modernization have largely continued on, even after the demise of formal, self-identified modernization theory. For example, Manuel Castells's description of "the Net and the Self" fits older categories of the modern and the traditional almost perfectly, with one vitally important change: for Castells, the values associated with the Net and the Self are associated with different groups *within* nations, and even to different aspects of an individual's daily life, not to societies as a whole.[35]

Modernization theory was more than just a passing fad, however. Innovations that it brought to the study of social change have altered the tradition. For example, while behavioral-systems thinking in the abstract Parsonian mode is not in vogue, some of its key concepts—feedback, dynamic equilibrium, formal understandings of interdependence, behavioral-functional analysis as one important component of the social analyst's arsenal, self-organization, the idea that relations structure systems or networks, a belief that model making and testing form the basic intellectual enterprise of the social scientist—are still with us.

Similarly, while thinkers about social change today do not treat nation-states as unified entities, the state nevertheless plays a much larger role in thinking about development than it did in prewar analyses. Even when scholars emphasize the role of nonstate actors, they do so by reference to the roles that states can—or ought—to perform. And while evolutionary thinking is pervasive once again, the idea of a convergence on certain common elements in modernity seems to many to fit today's post–Cold War world better than the time in which it came to prominence. (Francis Fukuyama's claim that history has "ended" is vastly exaggerated, but it is nonetheless true that even the communists are capitalists today, for better or for worse.)[36] There are differences, of course—modernity is now seen as producing inequality as often as it is seen as democratizing and egalitarian, for example—and the tone of analysis can range far more widely, from the triumphalism of a Fukuyama to the epic vision of crisis and change of Castells to the cold-eyed Marxian realism of David Harvey to the fierce anger (or despair) of Foucauldians or of dependency theorists.

If one takes Daniel Bell's landmark *The Coming of Post-Industrial Society* as one's exemplar of late modern analyses of modernization and development (it was the first great late-modern work on modernization), then the continuities with and changes from "high modern" modernization theory both stand in sharp relief: advances in the control of nature are still the great engines driving social change, and such advances are still tightly linked to the spread of instrumental reason in the form of science-based technology and of market-based economics. The specific engines powering development have changed, with information technologies and "intellectual technologies" replacing energy conversion devices as the chief engines of growth, a change with major implications for industrial policy. Nevertheless, change is wrought by the same instrumental, rational hand.[37]

There also remains in Bell's work, as in most late modern analyses, a sense of "stages" in economic growth, if not in modernization writ large. Bell's description of the arc of progressive economic development from agricultural to industrial to service-based economies has become so widely accepted as to be common knowledge. Finally, he continues to understand modernization as a shift to a world of constant change and disequilibrium. Responsibility for adaptation to this world of change is more highly distributed, however, with workers, governments, and firms all needing to reinvent themselves constantly in order to adapt to an ever more complex, ever more rapidly changing world.

There are significant departures in late modern social science from earlier analyses of social change, however. First, and foremost, where both prewar and early Cold War scholars saw modernization as a *tightly coupled* total change in society, late modern analysts from Bell on have seen the economic, political, and cultural realms as *loosely coupled*. That is, while they still understand modernization as a total change driven by changes in the economic-technical realm, there are, in the polity and culture, multiple possible adaptations to these changes. As Bell puts it, the econo-technical world "poses questions" for the polity and culture, but it does not determine the answers.[38]

As a result, late modern social science displays a newfound emphasis on divergence and choice rather than on convergence. This divergence still takes place within boundaries—not everything is on the table—but Asian-style state-led capitalist development, for example, is now seen by many as an alternative to, not a step toward, American individualist capitalism. With this shift toward divergence and choice has come a debate about the grand narrative: the "Rise of the West and westernization of the rest" storyline now must compete with the "clash of civilizations" and the "worldwide developmental project" and the "we have never been modern" storylines.[39]

Paradoxically, the rise of extremist, militant Islam and its link to Middle Eastern petro-despots; the ascent of non-Euro-American industrial powers, such as China, Japan, India, and Brazil; and the growing concern over environmental degradation have conspired to bring energy back to center stage, where it plays opposite information. This trend is most visible in the work of environmental historians, for whom the nineteenth century is once again the age of coal and steam and the twentieth the era of oil and the internal combustion engine. In such studies, energy available per capita and information available per capita bid to join income per capita as standards for measuring development. Whether this twenty-first-century version of modernity represents progress, or simply describes it once again, remains to be seen.

CHAPTER SIX

A Model Science?

Nowadays it is clear that a model rather than a definition serves to represent the complex variables of a complex situation, thing, or process. A model serves better to put together empirical descriptions economically and surely and to handle summarily things of many dimensions, little-known organization, diverse functions and processes, and intricate connections with other things. Definitions are too shallow and too full of verbal traps; summaries of propositions are too slow, piecemeal, and cumbersome. And certainly communities are such complex things.

Conrad Arensberg, 1955

Anthropologist Conrad Arensberg spoke for many postwar social scientists in his advocacy of models and modeling, and in linking modeling to the scientific study of organization, function, and process in a complex world. Such an approach to social science is so commonplace today that it can be hard to remember just how recently it came on the scene, and how significant a change it represented in the basic concepts, practices, and goals of social science.

Before 1950, social scientists *rarely* engaged in modeling. Then, between 1955 and 1970, modeling suddenly became a common practice among social scientists across a wide range of fields (figure 6.1). Unlike many concepts and practices that flourished during this peak of high modernism, however, modeling did not fade away. It continued to grow in importance in all fields, to the point that the typical article in the social sciences today (just as in the natural sciences) is oriented around the application, evaluation, or extension of a model.

As Nobel laureate economist Robert Solow argued in 1997, while postwar economics had been criticized for being excessively formalist, that was not quite accurate: in his view, economics became not so much a formalist as a "technical" discipline, with the focal point of its transformation being the rise

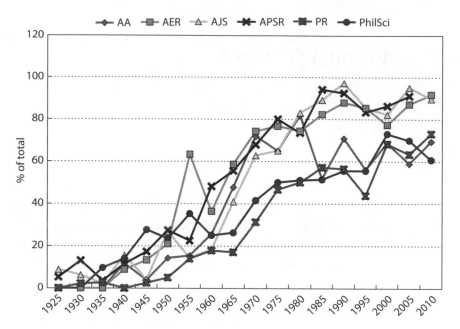

Figure 6.1. Articles in selected journals in which a form of the word *model* is used (not including nonrelevant usage). Journals are *American Anthropologist* (*AA*), *American Economic Review* (*AER*), *American Journal of Sociology* (*AJS*), *American Political Science Review* (*APSR*), *Philosophy of Science* (*PhilSci*), and *Psychological Review* (*PR*).

of "model-building." Such models are mathematical, which gives the impression to outsiders that they are formalistic and theory-driven, and, indeed, there is a "tendency for theory to outrun data." In truth, however, modelers are "obsessed with data," partly because models are judged by their "fit" to data, not by their fit to theory, and partly because "theory is cheap, and data are expensive."[1]

While economists pride themselves on being more technical than other social scientists, the same shift toward the technical could describe the work of a majority of practitioners of most of the social sciences today. Anthropology is the only field in which the exception (interpretive description and analysis, usually in terms of a theory, but not tethered to a model) is the rule, and there are more modelers in anthropology than one might expect.

In the social science journals surveyed (see introduction), before 1950 less than 7 percent of the articles used the word *model* (or its variants) *at all*, and very few of the articles that used the word used models as objects or tools of scientific inquiry—that is, with *very few* exceptions (perhaps 12 out of 1,230), the articles did not engage in the construction, elaboration, extension, testing, comparison, or other analysis of a model or models.[2] By the 2000s, however, over 90 percent of the articles in the *American Economic Review*, *American Journal of Sociology*, and *American Political Science Review*, and roughly 70 percent of articles in the *Psychological Review*, *Philosophy of Science*, and *American Anthropologist* used "model," with almost all of the articles that used the term using it multiple times (as opposed to the typical singleton use in pre-1950 articles).

In addition, in the 2000s, in the journals in which the term was used most often, it was, with very few exceptions, being used because models were the primary objects of scientific inquiry in those articles. In the other three journals, there was a wider range in how often models were primary objects of scientific inquiry, with between a third and a half (depending on the year) in the *American Anthropologist* and *Philosophy of Science* using models in that way, while virtually all of the articles in the *Psychological Review* that used *model* (about two-thirds of the articles) did so because they engaged in modeling as a core practice.

Modeling is so dominant a practice that some philosophers of science have begun to argue that late modern science is really about the construction and testing of models, not theories. In the late 1990s, Nancy Cartwright famously extended this insight beyond present science to argue that all science is an attempt to create partial models of a "dappled world."[3] Others (e.g., Bas van Fraassen) have argued that theories themselves are but families of models, though this version of the "semantic view" has its critics.[4]

In this later, post-1970s, period, not only has the number of models proliferated, so too have the types of models, along with the methods, modes, and applications of modeling. In his article "Models in Science" in the *Stanford Encyclopedia of Philosophy*, Stephen Hartmann (one of today's foremost authorities on modeling) compiles a fascinating list of the different types of models, including scale models, idealized models (Aristotelian, Galilean, and approximate), analogical models, phenomenological models, data models, toy models, and structural models (in the mathematical sense of a structure), and he notes that a wide variety of things can serve as models, from physical objects to fictional objects, set-theoretic structures, descriptions, and simulations.[5] In short, the practice of modeling has grown and differentiated rapidly and so has become fundamental to all late modern science, social and natural.

The rise and spread of modeling in the social sciences after World War II falls into two distinct periods, each raising its own questions. Why was there a sudden rise in modeling around 1955–70, when it had played so small a role before? And why has modeling undergone continued growth since the 1970s, when many of the concepts and practices that flourished during the age of system were cast aside?

In answer to the first question: the initial rise of modeling went hand in hand with the high modernist project for social science and so was an expression of the bureaucratic, high modern worldview, as well as a product of the continuing organizational revolution, its patrons, and its chief technologies, most notably, the digital computer. It was a product of the age of system and was championed by those who championed the rest of the high modern agenda.

Specifically, modeling involved the use of class of "manipulable mobiles" that served as ideational tools of remarkable power, both intellectually and culturally.[6] These manipulable mobiles satisfied high modernist demands for a new kind of "working knowledge," as Joel Isaac terms it, and they were perfectly suited for creating packageable research products of the size, scope, and form the new patrons and users of social science research found valuable.[7]

The answer to the second question has three parts. First, while much changed after the 1970s, not only do the characteristic problems social scientific models were developed to address still exist, but they have grown and differentiated. Modeling is still relatively effective in dealing with these problems. There is thus a strong practical reason for continuity amid the broad ideological fracture of the post-1970s era. Another way of putting this claim is that the "late modern" grows out of, modifies, and reinterprets the "high modern"; it does not emerge wholly new, without a past.

Second, modeling diversified rapidly, especially from the late 1970s onward. Thus, modeling became a universal method by becoming a universe of methods, a development that was enabled by a constellation of factors, some intellectual and some institutional. Perhaps the most important of these was that computing continued to grow and diversify, and the digital computer is the modeler's muse.

Third, as a practice oriented around the exchange and shared use of manipulable mobiles, modeling is a fundamentally instrumental, technical enterprise (though it can be a fantastically abstract one as well), and this instrumental, technical orientation has fit the institutional and intellectual ecology of the late modern period—not to mention its generally miserly regard for truths that cannot be monetized, materialized, or packaged.

Modeling's high modern evangelists—including Norbert Wiener, Karl Deutsch, and C. West Churchman—established a means of defining and employing models that helped redefine social science. These new, high modern approaches led to rapid growth and differentiation in the practice of modeling, to the point that some models have experienced "second lives" quite different from what their first creators intended. These new conceptions, however, are a logical consequence of models becoming not merely intellectual technologies but also intellectual commodities—manipulable *mobiles*—conceptual tools made for exchange.

Modeling's Many Meanings

Model is a word with many meanings, and those meanings have changed notably over the past century and a half. This poses an interesting challenge for the historian: is it preferable to define the term (perhaps with the assistance of today's philosophers of science), searching for when and how people began to use the term to mean something close to that definition? This would impose an admirable consistency on the search, but quite possibly at the expense of understanding the perspective of historical actors who interpreted the term in a different way. Or is it better to inventory all the uses of the term and all the debates about its meaning, locating them in context? That method would stay true to the language and concerns of historical actors, but quite possibly at the expense of understanding the history of the concepts and practices that concern us here.

These are not small choices. For example, in her interviews with practicing scientists, Donna Bailer-Jones finds a wide variety of definitions of the word *model*, some that make *model* and *theory* almost synonymous and some making them very distinct.[8] Similarly, Stephan Hartmann finds nearly two dozen different kinds of models within science.[9] Picking one (isomorphic structural analogies, say) as the true "model" and tracking its history instead of another (tangible physical mechanisms, say, or computer simulations) would lead to very different histories.

If one steps outside the world of the sciences, then the term's uses become broader still. One meets models and "supermodels" in the world of art and fashion; scale models, working models, test models, and this year's model (better than the last!) in the world of engineering; and, in the ever-present corporate parlance, business models. In the realm of the semipopular discourse of public intellectuals (and op-ed writers), one also encounters the corporate model, academic model, capitalist model, socialist model, single-payer model, fee-for-service model, and so on. None of these uses is technical, but some employers of the term clearly attempt to draw on the authority of science; are they to be included in this history?

Bailer-Jones opts for generality, stating, "A model is an interpretive description of a phenomenon that facilitates access to that phenomenon." She further distinguishes models from theories by noting that theories aim to be universal and so are not directly applicable to any particular empirical phenomenon, though they have the capacity to be so applied by being incorporated into models of that phenomenon.[10]

As interpretive descriptions that facilitate access, Bailer-Jones sees models as playing an essential role in science, one that neither theories nor empirical inductions from data sets play. Most recent writing on models in science agrees that models function as essential parts of the creative process in science and are not simply preliminary or incompletely worked-out theories. Margaret Morrison, for example, argues that models are "mediators" between theory and data, with such mediators being (at least partly) "autonomous agents."[11] Eric Winsberg concurs, emphasizing that this mediation is an active, constructive process.[12] To him, one cannot understand modeling unless one realizes that no complex system can be modeled simply by drawing down available theory, nor can it be modeled by empirical generalization alone. Rather, it involves theoretical deduction, empirical generalization, and a host of techniques specific to modeling that simplify calculation and generate a better "fit" of model to data, techniques that often have no basis in theory or in the physical system being modeled (they "save the appearances," as a semi-mythical Ptolemaic astronomer might say).

From another perspective, cognitive psychologists and philosophers informed by their work, such as Nancy Nersessian, hold that we think in terms of models, images, and symbols that represent other, larger, more complex things.[13] Models are thus simplified, yet essential, parts of the acts of mental representation and of generalization (an idea whose roots we saw in chapter 3). Likewise, many linguists and psychologists (and some philosophers) see strong analogies between models, metaphors, and paradigms, with metaphor and analogy being crucial to the creative extension of ideas from one domain to another.[14] Seen this way, models belong to the vast family of symbols and representations, a family whose relationships to the laws of logic on the one hand and to "objective" reality on the other were the subject of intense debate for most of the twentieth century. Thus, continental structuralism, as a revolutionary perspective on the nature of linguistic representation, and mathematical structuralism (seen in set theory and Bourbaki), as a revolutionary perspective on the nature of symbolic representation, were powerful parallel developments whose combination did much to shape the theory and practice of modeling.[15]

Hartmann, broadly speaking, agrees with this approach, trying to bring it more fully within the traditional corpus of philosophy of science. Following the cognitivists and linguists, he implicitly defines a model as a *representation*: "Models can perform two fundamentally different representational functions. On the one hand, a model can be a representation of a selected part of the world (the 'target system'). Depending on the nature of the target, such models are either models of phenomena or models of data. On the other hand, a model can represent a theory in the sense that it interprets the laws and axioms of that theory. These two notions are not mutually exclusive, as scientific models can be representations in both senses at the same time."[16] A theory, presumably, serves a different function than representation (explanation, say) and so can be differentiated from a model on functional grounds.

A history of modeling in science, then, would be a history of the emergence and spread of a particular form of representation (assuming that there are forms of representation that *are not* models, which seems safe to assume, given the radical shift in usage noted above). Following Bailer-Jones, we should remember that this form of representation is always interpretive, not simply driven by the phenomena, and, following Morrison and Winsberg, we should keep in mind that models are both products of and tools for particular kinds of thinking.

Suffice it to say then that a model is a *simplified representation* of something. Simplifying might involve scaling down (or up), reducing the number of variables or components, idealizing or abstracting from the messiness of concrete situations, making something new and strange more familiar through analogy or representation in a more familiar, tractable form, and more. Such simplified representations may be physical or mental, real or fictional, tangible or symbolic, formal or informal, qualitative or quantitative, in any combination.

As simplified representations, models serve many useful cognitive and practical functions, ranging from simple recall of complex information ("chunking" it as George Miller would say, which enables a hierarchic memory structure) to the extension of insights by analogy (which enables that hierarchic memory structure to be thoroughly associative as well). Models can be made to be easier to manipulate, mentally or physically, than the originals; to test the application of a theory to a concrete, real-world situation; to ease communication; or to direct attention or serve as a heuristic even when they are not providing isomorphic analogies.

Defined so broadly—as a simplified representation of something—models are everywhere, used by everyone everyday—a statement with which the leaders of the cognitive revolution would agree wholeheartedly! More specifically, however, a scientific model, in the most general sense, is a simplified representation of a natural phenomenon, within which ambit we include humans and their thoughts and actions, individually and in groups.

In science, models are useful when one is dealing with a complex phenomenon—that is when simplification is most valuable. Simplification always entails costs as well as benefits, so one would expect that modeling would become more useful as the phenomena under study become more complex. *Complex*, however, is not a simple word. What makes a phenomenon complex? The number of components in a system? How tightly they are coupled together in that system? The mode in which they interact—through feedback loops, say, as opposed to linear causal relations? The difficulty of representing the phenomenon in familiar terms, images, concepts, or symbols?

Take Isaac Newton and his *Principia* for example: on the one hand, Newton's theory of universal gravitation and his inverse-square law can be stated quite simply; on the other hand, he had to redefine mass, force, inertia, and causation; break down the distinctions between heavenly and terrestrial motion; redefine natural philosophy as rational mechanics; and invent the calculus in order to arrive at, explain, and justify his "simple" theory and equation. Making sense of Newton certainly was a complex endeavor then, as now, and his use of simple mechanical analogies or models (a planet is like a ball on a string, or like a cannonball eternally falling) clearly was intended to help simplify what was new and strange, and therefore complex. "Complex" phenomena might thus be better defined not in relation to their intrinsic "complicatedness" but in relation to their availability to the human mind in a particular time and place— and, in science, to the trained mind of the scientist in a particular field.

Manipulable Mobiles

Simplified representations have been part of science and natural philosophy for a long time. How, then, can such a definition of models be used to explain modeling's sudden rise as a distinctively new practice in the middle of the twentieth century?

To answer this question, let us turn from philosophy toward history, in particular, a history of the uses of models. This history reveals that while all models are simplified representations, not all simplified representations are models of the sort that were invented and embraced after World War II. A new conception of models emerged around midcentury, along with a new set of goals, conventions, techniques, technologies, audiences, and markets for those models. In this conception, models were understood as simplified representations of a particular type—what could be called "manipulable mobiles." Not simply useful pedagogical devices or shorthands for what was already known, they became essential tools for discovery. They not only produced experimentally valid knowledge but also were central to science writ large.

Mary Morgan offers a historically informed picture of models in *The World in the Model: How Economists Work and Think*. Her argument is simply stated: "From the late nineteenth century, economics gradually became a more technocratic, tool-based, science, using mathematics and statistics embedded in various kinds of analytic techniques. By the late twentieth century, economics had become heavily dependent on a set of reasoning tools that economists now call 'models': small mathematical, statistical, graphical, diagrammatic, and even physical objects that can be manipulated in various different ways . . . These manipulable objects are the practical starting point in economic research work."[17]

To Morgan, the invention and embrace of this new kind of scientific object "involved not just the adoption of new language of expression into economics (such as algebra or geometry), but also the introduction of a new way of reasoning." This new way of reasoning was the "scientific style" of "hypothetical modeling"—a style based on thought experiments.[18] In some ways it was very much like the early astronomy of Galileo, except that it was much more oriented around the use of certain shared intellectual "tools"—symbolic forms used to denote certain things (in Morgan's book, economic ideas) in *manipulable* ways. The possibility of manipulation is crucial, for "it is this possibility for manipulation that turns such pictures [as the *Tableau économique* of Quesnay or the Edgeworth Box] into models for the economist."[19]

Thus, Morgan's picture of a model is oriented around its use. To her, modeling is fundamentally an investigative activity, a kind of experimentation, and models serve that purpose by both "giving form to ideas and making them formally rule bound."[20] Their manipulability according to formal rules is crucial to their role as "tools or instruments of scientific investigation," for models are not merely inscribed, circulated, studied, or discussed, they are *used*: they are built, inspected, tested (sometimes against empirical data, sometimes against hypothetical conditions), tweaked, retested, rebuilt, and so on. In this they have much in common with modern technical drawings in that a set of formal, conventional rules specifies a mode of representation that allows those who understand that mode to take a particular representation, inspect it visually, and then build what it represents—or manipulate that representation and thus the thing to be built.

For models to serve this function, they must not only be *manipulable*, they also must be *mobile*; that is, they must have the characteristics Bruno Latour ascribed to certain new types of inscriptions he termed "immutable mobiles" in "Visualization and Cognition." Namely, they must be (1) mobile (movable over long distances); (2) unchanged in their meaningful characteristics when so moved; (3) flat; (4) scalable; (5) reproducible; (6) recombinable (as when maps of different sections of a coastline are joined); (7) superimposable (as when population data is added to a topographic map); (8) capable of being merged with written text; and (9) capable of being "merged with geometry" (they convert multiple dimensions and vast scales to two dimensions and convenient sizes for synoptic visual apprehension).[21]

One need not embrace Latour's version of actor-network theory or his account of the establishment and defense of facts through the mobilization of allies in an agonistic struggle in order to see that he is on to something enormously important here, something that helps us understand why the Renaissance invention of linear perspective and the later invention of projective geometry (both crucial for technical drawings); the eighteenth- and nineteenth-century production of increasingly complex maps of spatial (and, eventually, temporal) distributions; the development of a host of new measurement and inscription devices in the late nineteenth century; the creation of vast bureaucratic record-keeping, accounting, and statistical systems in the nineteenth and twentieth centuries; and the post–World War II conversion of all codeable information to digital form, for example, all should be linked not only to new devices but also to new conceptions of the world and its workings.[22] All of these, in Latour's terms, enabled the creation of new classes of immutable mobiles and so enabled new modes of visualization, new forms of understanding, and new plays of power.

As far as models are concerned, the least applicable aspect of the notion of an immutable mobile is the immutability of the inscription. A model is an immutable thing in the limited sense that the inscription that denotes it must be (close to) unchanged when moved. In addition, certain things about a model must remain the same across certain kinds of transformations (e.g., geometric relationships across reorientations in space). But if models were wholly immutable, then they would not be manipulable, testable against empirical experience, and combinable or superimposable, which is why they are useful tools for investigation.[23] Rather, models are mutable, *but in bounded, rule-governed ways*.

The rule-governed aspect of scientific models can vary with their intended functions. Some models are used as *heuristics* rather than as experimental probes; such models tend to be expressed in less formal, less rule-bound languages. There are still rules to the representation, nonetheless, that enable such a general model (heuristic though it may be) to be "played with" in ways that inspecting something that is not a model will not allow.

For example, in postwar cognitive psychology, the most general working model of the mind was that it was an information processor. This was such a broad and general model that it might be more accurate to call it a root metaphor, but whether model or metaphor, it directed attention to certain problems and framed questions in certain terms (in this case, in the language of information and its representation, storage, transformation, and communication). Successive versions of this general heuristic model were given more concrete forms through analogies to successive generations of information technologies—first to serial processors with "von Neumann architectures" (having five vital "organs": an input and an output device, a central processor, and separate short-term and long-term memory structures), and then to time-sharing systems with attention-directing operating systems, communications networks with a mix of hierarchic structures and associative links, evolving ecologies of cellular agents, and so on.

To Herbert Simon, Allen Newell, and other pioneers of cognitive psychology, these metaphorical models became scientific models once encoded in a formal language. The formal language that made such models intelligible, workable, and useful, was the *program*. The program, coupled with the digital computer on which it ran, made the information-processing model of mind into a new and powerful experimental tool, a manipulable mobile.

This bounded manipulability is the key to a scientific model's utility—and the source of its chief limitations. Without these rules, the manipulations teach us nothing, but so long as there are rules, the model only can teach us things that make sense within those rules. If the rules are wrong (if Riemannian geometry *does not* describe the physical universe, say) then the model will fail—or worse, mislead.

The rules that govern any particular model are partly the products of human convention, especially disciplinary convention, and partly the products of nature's constraints. Models thus lead a "double life," as Morgan puts it: *"models function both as objects to enquire into and as objects to enquire with."*[24] As such, they must conform to the rules or conventions governing the form of the model (the rules of calculus, projective geometry, numerical simulation, etc.), and they must offer points of potential correspondence to the world the model purports to represent (or denote) in simplified form. This will be a *mediated* correspondence, as the model never promises a perfect representation, but rather a simplified one, but if it has no such points of potential correspondence, it no longer functions as a model.

This "double life" gives models empirical, experimental utility, though of a limited sort—to Morgan, models may "surprise, but never confound" the scientist with their results.[25] It is also a potential source of tension, however, as many a model today is made to "save the appearances" and fit the relevant data through the use of assumptions and equations that can have no real-world analogs, while others threaten to detach from any connection to empirical reality in their endless pursuit of formal elegance.

Putting these pieces together, my model of high modern modeling thus assembles Latour's notion of immutable mobiles with Morgan's definition of a model as a manipulable denotation of an idea (created for the purpose of inquiry) and the cognitive scientist's understanding of models as simplified representations of natural phenomena. The result is a picture of a scientific model as a *manipulable mobile*: a mobile denotational device that is simultaneously a simplified representation of the world and a rule-governed instrument for inquiry.

While models of various sorts, including hypothetical models and calculational devices, have existed for a long time, what was new about high modernists and their approach to modeling was precisely this conception of a model as a manipulable mobile, paired with the belief that such models were not merely acceptable but fundamental to science. When one combined these ideas with the innovative tools and resources of the postwar era, one had both the seeds and the soil for a model harvest.

An alternate reading of this history would be that models always have been manipulable mobiles but that they only came to be understood as such in the postwar era, with that understanding changing their role in science. There are many historical parallels that make this interpretation plausible as well: for example, many devices throughout history have been energy-conversion machines, but they only came to be understood as such in the nineteenth century, and that new understanding led to changes in the design and use of such devices, with a more universal conception of energy leading to a more ubiquitous use of energy-conversion devices and the attempt to design machines that conserved all forms of energy, not just human labor.[26] Similarly, many technologies have been information or communications technologies, but they only came to be understood as such (that is, as members of a particular class of technologies with a unique role in society) in the second half of the twentieth century; this more universal understanding of those technologies led to changes in their design and use (e.g., universal interoperability through the digitization of everything being a deliberate design goal), and to a dramatic elevation in their sociocultural status.

Whether the mid-twentieth-century conception of models as manipulable mobiles was an invention, a discovery, or a reinterpretation, it was new, potent, and productive.

Modeling and High Modernism

High modernists may not have used the phrase *manipulable mobile*, but they did develop a new understanding of models and modeling. And, as with all aspects of high modern social science, this new understanding was not static. It did not emerge from Norbert Weiner's Jovian brow, full-grown; rather, it changed, even over the relatively short course of the high modern period. The first iterations of this model of modeling emphasized structural isomorphisms, formal descriptions, and fruitful analogies as the keys to good models, while versions from the mid- to late 1950s increasingly emphasized that models were simplified representations with instrumental utility.

For the most part, however, these were differences in emphasis, sibling conceptions in a close-knit family of ideas. All in the family could agree that a model was a simplified representation of a *system*, encoded in a formal language. Such formalized models could take the form of systems of differential equations, lines of code in a computer program, sequences of if-then statements proceeding from certain axioms, statistical tables and associated equations marking correlations among values, and more. Such models almost always focused on the system's basic *structure*, usually understood as a set of processes or relations. In addition, high modernists generally were fascinated by the act of representation itself, especially in symbolic form, and the nature of formal symbol systems.

They did not eschew looser usage entirely, sometimes using *model* to mean exemplar or version or any simplified representation, but it is clear that high modernists saw modeling as a scientific practice and that, to them, the core of that practice was the encoding of a model in a formal language, with the structure of the model in that language ideally being isomorphic with the structure of the thing being represented.[27] Herbert Simon often said that "the program *is* the theory."[28] By the 1970s, his colleagues would have agreed, only most of them would have substituted "model" for "theory" in that sentence.

Where did this much more specific, *structural* understanding of modeling—and dramatically more extensive and intensive use of modeling—come from? In other words, in what context shall we place the rise of modeling after World War II?

The contexts that make the rise of modeling intelligible are those associated with the bureaucratic, high modern worldview and the organizational revolution. Even if one leaves the fascination with formal modeling out of the definition of the high modern worldview, the intellectual commitments that made up that worldview—to system, structure, process, formalism in representation—were perfectly suited for the embrace of modeling. And the context that supported those commitments—the continuing organizational revolution, especially as embodied in the new patronage systems and communications and information technologies of postwar science, especially the digital computer—provided fertile ground for the rise of modeling.

Modeling and Modernist Practice

Two other key elements of the context for the rise of modeling were a self-awareness regarding the nature and meaning of representation and an uncertainty about the possibility of grounding truth in anything beyond convention, concerns that were broadly characteristic of modernist thinking in both scientific and literary culture. In art, these concerns translated into a fascination with style and form; in linguistics, the study of the structure of grammar and semiotics; in math, set theory and the development of symbolic logic; in natural and social science, the development of formal theory, modeling, simulation, and a persistent awareness of the necessary distance between the thing and its representation.[29]

All this still begs the question why should the experience of modernity lead to a new kind of focus on the act, nature, and structure of representation? This is a question that has fired the minds of cultural critics from the late 1800s to the present. Roland Barthes even went so far as to claim that since about 1850, the "whole of literature, from Flaubert to the present day, became the problematics of language." (Flaubert's famous statement, "What strikes me as beautiful, what I should like to do, is a book about nothing, a book without external attachments, which would hold itself together by itself through the internal force of its style," certainly helps Barthes's case.)[30]

To answer this question, let us begin with time and change. Dorothy Ross sees modernism in social science as something grounded in a shift in temporal or historical consciousness, with a Progressive Era generation of American social scientists embracing an evolutionary historical consciousness that put them at the forefront of a revolutionary evolutionary change.[31] To them, social-economic-technological evolution had produced a radically new world, a world of constant change. A later, post–World War I generation agreed that theirs was a world of ceaseless flux, but they increasingly located that world not at the leading edge of continuous evolutionary development but somewhere across the Rubicon of modernity.

How would this shift in historical consciousness affect attitudes toward science, nature, and the act of representation? Malcolm Bradbury and James McFarlane give us a clue in their classic work on modernist art and literature:

> But one feature that links the movements at the centre of sensibility we are discerning [modernism] is that they tend to see history or human life not as a sequence, or history not as an evolving logic; art and the urgent now strike obliquely across. Modernist works frequently tend to be ordered, then, not on the sequence of historical time or the evolving sequence of character, from history or story, as in realism and naturalism; they tend to work spatially or through layers of consciousness, working towards a logic of metaphor or form. The symbol or image itself . . . helps to impose the synchronicity which is one of the staples of the Modernist style.[32]

In other words, narrative and other story forms for representing events traditionally depend on an implicit chronological structure. If such chronological structures can no longer be taken for granted (and may not even be acceptable, to some), then other forms of representation must be found. Perhaps these will be "images," as Ezra Pound described them: "An image is that which presents an intellectual and emotional complex in an instant of time."[33] Or perhaps they will be symbols, systems of equations, tables, graphs, flow charts, programs, or, more generally, models. And perhaps that is why we almost never call modelers "authors" or "writers" of their models; that would link them to the world of narrative structure, which is a world they have left behind.

At the same time that the modern world is a world of change (so changed, it is separated from its past), it is also a world of interdependence, as Thomas Haskell argued over thirty years ago.[34] In a world of rails and wires, and national (and international) market economies, causation "receded," and the new, changed world appeared vastly more complex and impersonal. How to know such a world? The specialist could know but a piece of the whole; the generalist could not hope to know anything. The community of professional social scientific experts, sharing a common language and knowledge base, could pool their specialized expertise and so solve the problems of our changing, interdependent world. They would validate this knowledge and their new status, as Ted Porter and Andrew Abbott each have shown, by developing a set of new techniques of measurement and quantification and of abstraction to universal principles.[35] And as the idea of models as manipulable mobiles suggests, they would develop a common set of intellectual tools, ones requiring specialized training to create, understand, and manipulate, fully available only to the "community of the competent," which could bring together the very different kinds of high value attached to abstract knowledge and practical utility.

In addition to these intellectual tools, another class of tool also appeared at this time, the media technologies of the late nineteenth and early twentieth centuries: photographs, phonographs, telegraphs, telephones, mass print, mass circulation color magazines, radio, motion pictures, and television.[36] While it is difficult to tie any one of these mass analog media to a revolution in consciousness (though one could make pretty good arguments for the telegraph and radio/television), collectively they constituted a series of radical challenges to, and opportunities for play with, conventional notions of representation, communication, and audience—and even of time and space. They enabled the capture of previously evanescent and fleeting moments in time, their manipulation in the studio, and their reproduction and distribution across vast distances to new audiences.[37] They were also, as noted earlier, quintessential products of the organizational revolution.

Some connections between these new technologies and the advent of modeling are quite direct (e.g., telephone and radio communications research led to the development of a reframed notion of feedback and of Claude Shannon's new information theory, which became the basis of a great many communications models in the 1950s). In other cases, the connections are more indirect and "Zeitgeisty" (e.g., the experience of seeing multiple, different photographs of the same scene—many of them retouched and manipulated—raised questions about what "truth" really meant in a representation, at least for some scholars).[38]

Perhaps most important of all, the modern world is not only a world of change and interdependence, it is a world of daily encounters with diversities and scales that threaten an end to every order.[39] The world of the model, however, is one that can be ordered and controlled. It can be made simple, uniform, *clean*. It can be resized, rescaled to suit our needs and desires. If the model no longer fits reality, then perhaps reality can be changed. Thus, modeling was a practice that fit well with modernism's distinctive historical consciousness, its preoccupation with representation, its focus on interdependencies, and its search for order.

Not only did modeling fit well with these intellectual concerns, it was well suited to the new modes of research funding of the mid-twentieth century. A social science oriented toward the interests of extra-university patrons, particularly military patrons, tended to have an instrumental bent, and a style of research that produced packaged "models" as products that could be tailored to that patron's particular needs or repackaged and "resold" without starting from scratch increasingly appeared to be the best exemplar of "productive" work. Such an approach to research produced results, articles, experiments, and even technologies—everything that looked like "research productivity" to research managers. Theories, in contrast, are harder to bind up and package as products—and still harder to complete. In short, models were the perfect *iterative* intellectual products for a world that rapidly came to accept, and even to anticipate eagerly, release 3.1.4.

The construction and testing of models was a natural fit for a science increasingly oriented around the production of articles and article-length technical reports by research teams, as opposed to the prewar style of social science in which being a productive scholar generally meant being an individual who wrote original books. Model builders, like experimentalists, could crank out the publications in impressive quantities once their basic "model/lab" was set up: Herbert Simon was an extraordinary case, as he regularly worked 100 hour weeks, but producing over 800 articles—more than twenty per year in his prime—was not just about working incredibly hard, it was about setting up a research machine primed to build, manipulate, and test model after model.

The Making of Modeling

Many of the central figures in the articulation of the new "model of man," *homo adaptivus* (discussed in chapter 3) and the new models of rational choice (discussed in chapter 4) also were key figures in developing a new vision of science in which modeling provided the perfect synthesis of theory and experiment, abstraction and application. C. West Churchman, for example, not only devoted the bulk of his landmark *Introduction to Operations Research* to ideas and practices associated with modeling, but he also was editor of the journal *Philosophy of Science* from 1948 to 1957, when analysis of science as a model-based enterprise first became a commonplace in that journal.

Rosenblueth and Wiener

The place to begin an examination of high modernism and modeling is not with Churchman, however. Rather, it is with Arturo Rosenblueth and Norbert Wiener, famed as co-authors (with Julian Bigelow) for their 1943 article, "Behavior, Purpose, and Teleology," which introduced many scientists (social and natural) to the concept of feedback via the model of the servomechanism.[40] Wiener later became even more renowned for his landmark texts on "cybernetics" (he coined the term), in which he developed new methods to model complex systems in terms of communications and control. An important piece that came in between these landmark publications, one sometimes forgotten in the cybernetic frenzy, was "The Role of Models in Science," first published in 1945.[41]

According to Rosenblueth (a physiologist) and Wiener (a mathematician with a statistical bent), science is fundamentally an instrumental enterprise: "The intention and the result of a scientific inquiry is to obtain an understanding and a control of some part of the universe." But this instrumental task is a difficult one and cannot be fulfilled with mere observation and description. "No substantial part of the universe is so simple that it can be grasped and controlled without abstraction." From abstraction, the step to modeling is swift: "Abstraction consists in replacing the part of the universe under consideration by a model of similar but simpler structure. Models, formal or intellectual on the one hand, or material on the other, are thus a central necessity of scientific procedure."[42]

The authors make a distinction between "material and formal or intellectual models." A "material model is the representation of a complex system by a system which is assumed simpler and which is also assumed to have some properties similar to those selected for study in the original complex system. A formal model is a symbolic assertion in logical terms of an idealized relatively simple situation sharing the structural properties of the original factual system."[43] In both cases, structural parallelism is essential for model making to be useful.

Next, they examine "the results of carrying model-making to the limit. Consider first material models. They start by being rough approximations, surrogates for the real facts studied. Let the model approach asymptotically the complexity of the original situation. It will tend to become identical with that original system. As a limit it will become that system itself. That is, in a specific example, the best material model for a cat is another, or preferably the same cat." The point of this exercise in carrying the model to the limit, of course, is not to show that the only good model for a cat is a cat (and still less the same cat). Rather, it is to show that modeling *inevitably* simplifies the world. In keeping with the emphasis on the human mind as a finite thing in an infinite world (a view, as seen in chapter 3, common to high modern social science), they conclude in the same vein: "Partial models, imperfect as they may be, are the only means developed by science for understanding the universe. This statement does not imply an attitude of defeatism but the recognition that the main tool of science is the human mind and that the human mind is finite."[44] Equipped with the right models, however, that finite mind could "grasp and control" an ever-larger slice of the universe.

Rosenblueth and Wiener's "The Role of Models" was widely read and widely cited, though their article "Behavior, Purpose, and Teleology" probably did even more to spread the gospel of modeling by providing a startling new model for human behavior, the servomechanism, which many social scientists immediately saw as revolutionary. Take F. S. C. Northrop's widely cited 1948 article "Neurological and Behavioristic Psychological Basis of the Ordering of Society by Means of Ideas." (Note that by *ideas*, Northrop means *symbols*.) In this piece, Northrop linked Rosenblueth, Wiener, and Bigelow's model of man as servomechanism to McCulloch and Pitts's model of the brain as a "neural net," seeing the union as something of "revolutionary significance for natural science, moral as well as natural philosophy."[45] Northrop is admirably clear, if somewhat breathless:

> Recent investigations by Warren S. McCulloch and Walter Pitts show not merely that certain biological organisms, because of the character of the neuron nets in their nervous systems, must know universals, responding to symbols as their exemplars rather than as mere particulars. Other investigations by Arturo Rosenblueth, Norbert Wiener, and Julian Bigelow show that not merely a human being but also robots with inverse or negative feedback mechanisms have purposes that define their behavior. When this purpose can be determined by information, such robots are called servomechanisms. In other words, the basic premise of both the traditional philosophical dualists and idealists and the traditional, supposedly scientific naturalists and mechanists, to the effect that natural and biological systems can have neither knowledge of universals nor normatively defined and behavior-controlling purposes, must be rejected.[46]

That is a big claim—these new models of mind and man have enabled us to supersede centuries of philosophical and scientific dispute. Even the nature-nurture, culture-versus-biology debate now can be solved:

> Cultural factors are related to biological factors in social institutions by the biologically defined purposeful behavior of human neurological systems containing negative feedback mechanisms and the normative social theory defined in terms of the universals which are the epistemic correlates of trains of impulses in neural nets that are reverberating circuits. Because overt behavior can be tripped by impulses from reverberating circuits whose activity conforms to universals, as well as by impulses coming immediately from an external particular event, the behavior of men can be, and is, causally determined by embodiments of ideas as well as by particular environmental facts . . . In short, in any culture embodied ideas defining purposes or ideals really matter.[47]

Mechanism, Organism, and Society

These grand claims depended on the acceptance of models as valid modes of reasoning. In his article, Northrop largely assumes their acceptance. Not so for Karl Deutsch, whose 1951 article "Mechanism, Organism and Society: Some Models in Natural and Social Science" offers perhaps the most perfect assemblage of high modernist ideas of the early 1950s. It is such a perfect exemplar because Deutsch walks through every step of the logical, the cognitive, the philosophical, and the historical justification for the use of models in science as part of his argument for embracing the new, cybernetic model of man.

The article begins with a simple declarative: "Men think in terms of models." Deutsch explains this claim as follows:

> Their sense organs abstract the events which touch them; their memories store traces of these events as coded symbols; and they may recall them according to patterns which they learned earlier, or recombine them in patterns that are new. In all this, we may think of our thought as consisting of symbols which are put in relations or sequences according to operating rules. Both symbols and operating rules are acquired, in part directly from interaction with the outside world, and in part from elaboration of this material through internal recombination. Together, a set of symbols and a set of rules may constitute what we may call a calculus, a logic, a game or a model. Whatever we call it, it will have some structure, i.e., some pattern of distribution of relative discontinuities, and some "laws" of operation . . . If this pattern and these laws resemble, to any relevant extent, any particular situation or class of situations in the outside world, then, to that extent, these outside situations can be "understood," i.e., predicted—and perhaps even controlled—with the aid of this model.[48]

As was common among high modernists, Deutsch took pains to note that the symbolic world was real. "In one sense, all these models are physical. They consist of symbols which are states of physical objects, and traces of physical processes, whether in brain cells, ink marks, electric charges, or what not. Similarly, the operating rules, according to which these symbols are to be permutated or combined, and new symbols derived from them, are constraints on physical processes . . . In this sense, *knowledge is physical process*, or rather a particular configuration of physical processes."[49]

He connects the physical world to the worlds of structure, knowledge, and information: "Knowledge in general depends, therefore, on physical structure. If anything is to be knowable by any physical process, there must be in it some unevenness of distribution . . . Unevenness, structure, distribution are fundamental physical properties of everything—all matter, all energy, all processes—in the universe we know, and even in any universe we can imagine. Indeed this unevenness is the physical condition of all knowledge, all observation and all representation by symbols, all imagination and all understanding . . . What interacts, has structure. And what has structure, can be known" (231).

Deutsch cites Rosenblueth and Wiener on models in science, especially material models, approvingly, noting, "In [their] view, every material model implies a formal model behind it" (232). From this, he concludes that "what may count in intellectual history, then, may be not only the actual properties of a physical or social process which people accept at some time as a material model for some other process, but rather the idealized or implied properties which they ascribe to the implied formal model behind it . . . In this manner, the Egyptian pyramid, with its rigorous order of a very few stones at its apex and the many stones bearing all the burden at the bottom, has served as a model for the conception of a 'hierarchy,' whether of priests or army officers, or of ideas, values, and purposes, such as in Aristotelian philosophy" (233).

Continuing this move into the history of science, he argues, "It is only with the development of far more complex operations toward the end of the Middle Ages, that we find mechanical models in greater complexity, slightly less inadequate for describing the world around us. Mechanisms can be taken apart and reassembled. This is crucial for the new models" (233). Indeed,

> The classical concept or model of mechanism implied the notion of a whole which was completely equal to the sum of its parts; which could be run in reverse; and which would behave in exactly identical fashion no matter how often those parts were disassembled and put together again, and irrespective of the sequence in which the disassembling or re-assembling would take place. It implied consequently the notion that the parts were never significantly modified by each other, nor by their own past, and that each part once placed into its appropriate position with its appropriate momentum, would stay exactly there and continue to fulfill its completely and uniquely determined function. (234)

Following Whitehead and Burtt, Deutsch states, "This classical notion of mechanism was a strictly metaphysical concept. No thing completely fulfilling these conditions has ever been on land, or sea, or even, as our cosmologists have told us, among the stars" (234). But even more important than this classical mechanical model's metaphysical nature was its limited scope:

> According to this classical view, an "organism" is unanalyzable, at least in part. It cannot be taken apart and put together again without damage. As Wordsworth put it, "We murder to dissect." The parts of a classical organism, insofar as they can be identified at all, not only retain the functions which they have been assigned but in fact cannot be put to any other functions (except within narrow limits of "de-differentiation" which were often ignored), without destroying the

organism. The classical organism's behavior is irreversible. It has a significant past and a history—two things which the classical mechanism lacks—but it is only half historical because it was believed to follow its own peculiar "organic law" which governs its birth, maturity, and death, and which cannot be analyzed in terms of clearly identifiable "mechanical" causes. (236)

This classical concept of organism would seem to be a natural fit in biology and in the social sciences, but "attempts have been frequent to apply this classical concept of organism to biology and to human society. On the whole they have been unsuccessful . . . Neither mechanism nor organism, in their classical forms, could explain well the peculiar social cohesion found in many societies, cultures or peoples . . . Neither of these classical concepts could easily be used to deal with the process of learning. Nor were they easy to apply to the problem of knowledge; nor to the way qualitative judgments are made; nor to predictions in many social situations, or for longer trends in history; nor to the problems dealt with by esthetics, ethics, or religion" (236–37).

The incapacity of either approach, mechanism or organism, to grasp the complexities of the world—or of our human ability to make sense of it—was, for Deutsch, one of the root causes of the great intellectual problems of the modern world: "Perhaps we may yet come to recognize that the deep cleavage between the 'natural' and the 'social' sciences; between 'reason' and 'intuition,' or 'reason' and 'wisdom'; between the search for the truth and the search for goodness—perhaps we may come to recognize how much these cleavages were amplified and exaggerated in the thoughts of many good men between the times of Galileo and those of Einstein, by particular historical and social conditions in the Western world, and perhaps by the rather unwieldy intellectual equipment available during that period" (237).

Awkward intellectual equipment had left us with partial knowledges incapable of being interconnected, and so left us divided. But what kind of equipment would be necessary to build the proper connections? In answer, Deutsch observes that "what all these notions were trying to describe were processes of organization: self-sustaining or self-controlling or self-enlarging or self-transforming processes, as the case might be. Yet the only model they used to describe these processes was human society itself, changing throughout history, of infinite complexity, and baffling to those who tried to understand it while participating in its conflicts" (238).

His hope is that "by continuing to make such equipment which fulfills these functions of communication, organization, and control, we cannot help in the long run but gain significant opportunities for a clearer understanding of those functions themselves" (239). By making new communications and control devices, new techniques and technologies of organization, we create new models for understanding ourselves and our world, even as we transform that world.

> The operations of communication and control have been performed for thousands of years. But these operations were largely carried on inside the nerve systems of human bodies. They were inaccessible to direct observation or analysis. They could be neither taken apart nor reassembled. In the new electronic machines of communication and control all these things can be done. Messages or control operations can be taken apart, studied step by step and recombined to more efficient patterns . . . The science of communication and control, which can be derived from this technology, and which Norbert Wiener has called *Cybernetics*, is therefore a new science about an old subject. It investigates the old subject of communication and control, but it uses the facilities of modern technology in order to attempt to map out step by step the sequence of actual events involved in getting these operations accomplished. (239–40)

"The test of the usefulness of this new science, as that of any science, must be its results," he continues. "In the field of scientific theory it must offer new concepts rather than mere explanations" (240). To this end, he develops a "generalized concept of a self-modifying communications network or 'learning net.' Such a 'learning net' would be any system characterized by a relevant degree of organization, communication, and control, regardless of the particular processes by which its messages are transmitted and its functions carried out-whether by words between individuals in a social organization, or by nerve cells and hormones in a living body, or by electric signals in an electronic device" (240).

"What are some of the notions and concepts that can be derived from this technology? Perhaps the most important is the notion of information" (241). Here Deutsch draws on Warren Weaver's exposition of Claude Shannon's information theory.

> Power engineering transfers amounts of electric energy; communications engineering transfers information. It does not transfer events; it transfers a patterned relationship between events . . . These patterns of information can be measured

in quantitative terms, described in mathematical language, analyzed by science, and transmitted or operated on a practical industrial scale . . . This development is significant for wide fields of natural and social science. Information is indeed the "stuff dreams are made of." Yet it can be transmitted, recorded, analyzed, and measured. Whatever we may call it, information, pattern, form, Gestalt, state description, distribution function, or negative entropy, it has become accessible to the treatment of science. It differs from the "matter" and "energy" of nineteenth century mechanical materialism in that it cannot be described adequately by their conservation laws . . . But it also differs, if not more so, from the "idea" or "idealistic" or metaphysical philosophies, in that it is based on physical processes during every single moment of its existence, and that it can and must be dealt with by physical methods. It has material reality. It can be measured and counted. (241–43)

Deutsch adds feedback as another vital tool for constructing models of complex social systems. "Another group of notions which has been growing in different sciences is that of typical pathways or sequences which must be recognized as wholes since their performance cannot be predicted from any single step or element contained in them, but only from all of them in conjunction . . . With the aid of these models, we may recognize a basic pattern which minds, societies, and self-modifying communications networks have in common. Engineers have called this pattern the 'feed-back'" (245).

From these concepts of "information, feedback, learning, and purpose . . . further concepts may be derived . . ." (248). Linked, and scaled up, these concepts can offer a model for a functional, adaptable humanity. "A learning net functions as a society, in this view, to the extent that its constituent physical parts are capable of regrouping themselves into new patterns of activity in response to changes in the net's surroundings, or in response to the internally accumulating results of their own or the net's past" (250). That is, they are adaptive—this is a model of homo adaptivus on a large scale. But if homo adaptivus can be scaled up and so become socially heterogeneous in its composition, can it not be transposed not up or down but sideways, into the machine?

Already modern calculating machines involve a balance between subassemblies, permanently constructed for specific purposes, and transitory subassemblies put together from general elements to serve temporary needs . . . The twin tests by which we can tell a society from an organism or a machine, on this showing, would be the freedom of its parts to regroup themselves; and the nature of the regroupings, which must imply new coherent patterns of activity—in contrast to the mere wearing of a machine or the aging of an organism, which are marked by relatively few degrees of freedom and by the gradual disappearance of coherent patterns of activity. (250–51)

Deutsch's article, like his later landmark works of political science, *Nationalism and Social Communication* and *The Nerves of Government,* embodies the full high modernist argument: models are central to thought because to think, we must simplify the world, and models are simplified representations of it. Progress in science, therefore, largely consists in building new models that are better, more fruitful representations of the world. These representations never can be perfect or total, so they are to be judged on their utility in bringing more of the world under our "grasp and control." Finally, in complex electronic machines, we have models for a vast class of complex phenomena that neither classical mechanical nor classical organismic models proved useful in framing, predicting, or controlling. Why are these machines such useful models? Because they make tangible, testable, and visible the otherwise inaccessible, invisible linkages that cause organized systems to be organized and systemic. They enable us to represent not just energy (which, as the capacity to do work, could be modeled crudely in mechanical terms) but information and its flows, which, as a measure of organization itself, is the sine qua non of system.

Deutsch was ahead of the modeling curve, but others soon followed him. Indeed, by the mid-1950s, it was not unusual to find a mathematician arguing that "any given type of number, any number system, is either a model made to fit a set of experiential data, (and the numbers in a given model are then assigned properties suitable to that data by being 'subjected' to properly-chosen postulates,) or else it is a model created out of intellectual curiosity, in the broad sense, and usually by deliberate analogy with the models of the first type."[50]

Other Models of Modeling

Similarly, C. West Churchman organizes his *Introduction to Operations Research* around the core task of the construction of a *"communications (or control) model of the organization."* Citing Deutsch (and Rosenblueth and Wiener) on the role of models in science, Churchman writes, "It is worth noting here, however, that a model is a miniature of, or compact representation of, an original. Usually models represent relevant points of interest in the original; these points can be combined so that the structure of the model and that of the original are similar." He is most interested in a communications model—a model of the intangible, invisible stuff that makes an organization an organized thing: "Management, men, machines, and materials constitute a system only by virtue of *organization.*" The communications model enables one to understand this invisible stuff that binds by helping one picture it: "A communication model is not mathematical; it is not used for accurate predictions or calculations. It generally takes the form of a diagram. Such a diagram enables one to bring together, from various fields of research, knowledge about organizations." This model can be seen as a "glorified kind of fish net, spider's web, or network of nerves through which 'information' passes or flows."[51]

Churchman devotes the bulk of the text to the details of the construction of this communications model. Part III of the book is devoted to "The Model" and consists of one huge chapter, titled "Construction and Solution of the Model." In it, he returns to the notion of models as simplified representations, not isomorphic analogs: "The model, it will be seen, is a representation of the system under study, a representation which lends itself to use in predicting the effect on the system's effectiveness of possible changes in the system. Of the three types of models to be considered, the iconic, analogue, and symbolic, the latter is of particular importance . . . Viewed generically, a scientific model is a representation of some subject of inquiry (such as objects, events, processes, systems) and is used for purposes of prediction and control. The primary function of a scientific model is explanatory rather than descriptive."[52]

Models, to Churchman, are useful in determining what effects changes in one part of the system will have on other parts of the system, in part because it is simply easier to manipulate the model than the real world. The utility of models, however, goes deeper than that: "But the importance of models to science is out of all proportion to even these obvious and massive advantages. In fact, since scientific theorizing itself becomes *identical* with model construction in some aspects, it follows that science would be as impossible in the absence of models as it would be in the absence of theory."[53]

In *A Study of Thinking* by Jerome Bruner et al., another classic from that annus mirabilis 1956, the notion of a "category" is virtually identical to that of a model, at least as used by the authors' fellow high modernists: it is something that reduces the complexity of the environment, enabling the organism to order and relate classes of events and thus to provide direction for instrumental activity. In the same vein, modeling is central to the core ideas of Herbert Simon. In his famous phrasing, humans are "boundedly rational": "For the first consequence of the principle of bounded rationality is that the intended rationality of an actor requires him to construct a simplified model of the real situation in order to deal with it. He behaves rationally with respect to this model, and such behavior is not even approximately optimal with respect to the real world. To predict his behavior, we must understand the way in which this simplified model is constructed, and its construction will certainly be related to his psychological properties as a perceiving, thinking, and learning animal."[54] Simon certainly took his own principle to heart, titling some of his most important works *Models of Man, Models of Discovery, Models of Thought,* and (in a fittingly reflexive move) *Models of My Life.*

Nor was Simon alone among high modernists in using the model to expound his ideas. A. F. C. Wallace wrote that the "mazeway" is best understood "as a model of the cell-body-personality-nature-culture-society system or field, organized by the individual's own experience."[55] Warren Weaver was a major contributor to and exponent of the Shannon-Weaver model of a communications system. J. C. R. Licklider and Robert Taylor similarly thought of thinking and of communications in terms of models and modeling in their crucial work that led to interactive computing and computer networking.[56]

Indeed, one of the only high modernists who persisted in thinking of his work as theorizing rather than as model building was Talcott Parsons, which

Figure 6.2. The link between modeling and communication. When mental models are dissimilar, the achievement of communication might be signaled by changes in the structure of one of the models, or both of them. J. C. R. Licklider and Robert Taylor, "The Computer as a Communication Device," *Science and Technology* 76 (1968): 21–31, 24. Image credit: Roland B. Wilson.

is another reason why Parsons was a bit out of step with his cohort, despite his embrace of systems theory.

Modeling's Second Life

Even as the particular set of meanings attached to models, and the ways that high modernists legitimated them, began to lose their hold during the 1970s, the practice of modeling continued to grow. And spread. And change. How, then, did a practice so deeply embedded in the high modern not only survive but flourish in the late modern?

First, there were important continuities between the late modern and the high modern periods, as well as important changes. Second, modeling diversified rapidly from the late 1970s onward, allowing it to be many things to many people. Third, modeling's instrumental, technical orientation fit well with the institutional ecology and general temper of the late modern research university (and broader culture).

Continuity and Change

Intellectually, the *modernist* period (roughly the 1880s to the 1920s) was fundamentally oriented around the discovery of modernity and the consequent attempt to create an instrumental social science appropriate to a world of change, interdependence, and diverse subjectivities. Almost all subsequent social science, whether "high" or "late," is modernist in this broad sense. Thus, one should expect significant continuities as one moves from one of these overlapping phases to the next.

The *high modernist* era was oriented around the concept of system and the consequent attempt to create a social science that could map and manage the structures, processes, and functions of systems, usually by tracing the flows of information, communication, or control in them. In this, it was consciously an extension, elaboration, and formalization of the modernist style, though the implications of such an extension often were revolutionary.

The *late modernist* mode, in contrast, came into existence as a challenge to high modernism (and by extension, to the ultimate assumptions and conditions of modernity), though it could not escape its ancestry as a product of those assumptions and conditions. It was fundamentally oriented around the closely related concepts of complexity, contingency, and choice, and the result was a grand variety of methods, concepts, tools, and techniques in which the parameters are the object of study as much as (or more than) the variables, the bounds even more interesting than that which they contain. In the late modern view, everything is liminal, permeable, networked, and negotiated.

The late modern period finds the social sciences (like so many fields) increasingly transformed into a set of technical subfields oriented around the shared use and exchange of a set of "tools" used to solve, manage, or (at least) represent certain types of problems. This transformation is in part an adaptation to the institutional ecology of the late modern research university, its primary patrons, and its corporate surround. It is also a response to the general intellectual-political climate of the "age of fracture," as Daniel Rodgers terms the late modern era.[57]

Late modern social scientists did not unlearn the challenges of the twentieth century for philosophy—they still sought a broadly conventionalist yet rigorously empiricist (and, ideally, experimentalist) methodology for producing knowledge of complex phenomena that are extraordinarily difficult to observe and control directly. There is thus a strong philosophical-functional reason for continuity amid the broad ideological fracture of the post-1970s era. The "late modern" grows out of, modifies, and reinterprets the "high modern"; it does not emerge wholly new, without a past.

At the same time, the models that have become so widely used are different than those that first entered the social sciences in the 1950s and 1960s.

Of the Bayesian, evolutionary, Prisoner's Dilemma, and input-output families of models, only the latter is closely associated with a high modern, systems approach, and only it follows the pattern of a swift rise to a plateau in the early 1970s followed by a slow decline. The other three classes of models have roots in high modernist tools and techniques but were much more easily adapted to fit late modern precepts. Bayesian models are all about the contingent reevaluation of evidence. Rational-choice theorists continue to employ the Prisoner's Dilemma, often picturing the chooser as atomistic and relatively unbounded (any faults in the chooser's capacities can be worked around through a grand "as if," with the environment's remorselessly rational selection mechanisms filling in for the individual's limited capacities wherever necessary), a stark contrast to the homo adaptivus of the high modernists (see chapter 3). Evolutionary models, generally, are models of selection without deliberate choice and of endless branching and variation, not of development and convergence (as discussed in chapter 5). All three of these types of modeling have grown explosively, while more explicitly high modernist forms have plateaued or declined but not grown.

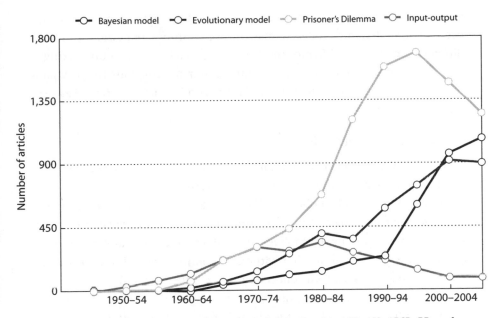

Figure 6.3. Articles employing common models in the *AA, AER, AJS, APSR, PR,* and *Philosophy of Science.*

In addition, many models have led second lives quite distinct from their first, a natural consequence of their being objects of exchange. Manipulable mobiles are intellectual commodities. They are made to be exchanged and circulated, and once an object circulates widely, its author begins to lose control over how it will be used, read, and interpreted. The author may lose control over even more basic elements of the product, such as which chapters to include (for a book, say), which illustrations to include (and where), or how the elements are to be arranged, priced, packaged, and distributed.

This same thing happens when scientific ideas are transformed into denotative devices as well as printed texts: the "Maxwell's equations" we use are really James Clerk Maxwell's equations as rewritten by Oliver Lodge (and then made easy to typeset and illustrate in textbooks); Francis Edgeworth's box is not exactly what he drew anymore; Anatol Rappoport and Kenneth Boulding used the Prisoner's Dilemma in their peace research to argue for the importance of just the sorts of collaboration among the "prisoners" that others have used the scenario to exclude; and game theory has expanded its empire from poker to economics to geopolitical strategy to the local politics of coalition building to evolutionary survival strategies on all sorts of scales to artificial intelligence research. The minimax equation has traveled along with game theory and been incorporated into models of many different things, but even it can have different meanings in different models: what, exactly, is one minimaxing—utility, money, opportunity, regret, freedom, power, information? That is a question with important consequences, and John Von Neumann's original answer may not be yours.

A Universe of Methods

Modeling diversified rapidly, especially from the 1980s onward, evolving into *a universal method by becoming a universe of methods*. On one level, if modern models are manipulable mobiles and thus are a species of "virtual machines" or "symbolic technologies," this diversification needs little explanation: having the right tool for the right job makes all the difference, as anyone who has attempted to change an oil filter without an oil filter wrench (or tried to assemble an Ikea bookshelf without the right size Allen wrench) can attest.

On another level, however, this answer is unsatisfying, for it takes as givens the very things that need to be explained: namely, the explosive growth of modeling, the availability of the "parts" needed to make the "tools" of modeling, and the acceptance of the legitimacy of models in many areas where such hypothetical or analogical tools had been viewed—in living memory—with considerable suspicion, if not hostility. So what, then, could explain the benign regress wherein growth enabled diversification enabled growth (and so on)? Where did the "parts" come from? Why were they so widely available, and why was modeling accepted in so many new areas?

The answers, as always, are multiple, involving developments on different levels in different arenas. Among these were a general postwar boom in mathematics, especially applied mathematics, with remarkable advances taking place in set theory, group theory, algorithmic analysis, symbolic logic, graph theory, the theory of mathematical structures, numerical methods, and in a host of statistical subfields. This series of mathematical revolutions generated an array of powerful tools relevant to modeling and a growing cadre of people trained in their use and confident in their utility. Gerard Debreu's work in economic modeling in the 1950s and 1960s, for example, was a direct application of the structuralist mathematical methods of Bourbaki to economics, and certain pieces of mathematical heavy weaponry, such as Brouwer's fixed point theorem, continue to show up with surprising frequency in social scientific models, especially those of some kind of equilibrium system.

In addition, the postwar era participated fully in the twentieth century's ongoing, accelerating data explosion. As a result, while data are still relatively expensive and theory is still relatively cheap, postwar modelers had access to ever wider arrays of ever larger and ever more diverse data sets. By the late modern period, those data sets increasingly became combinable and superimposable (to use Latour's terms) due to their reduction to a common digital format and the interlinking of digital databases. Say's Law, which says that supply creates its own demand, may not be strictly true for science (or in general, for that matter), but it is hard to escape the feeling that these piles of targeted data called specific tools into existence in order to deal with them.

These successive innovations in mathematics and the data explosion meant that modelers increasingly had ready stocks of well-made, reliable parts from which to assemble their models—and a reasonable assurance that a relevant mathematical tool or data set would become available in the near future, just in time for them to use it in their next grant application, publication, or annual report.

Models and Computers

In addition, as a practice oriented around the exchange and shared use of manipulable mobiles, modeling is a fundamentally instrumental, technical enterprise (though it can be a fantastically abstract one at times as well), and this instrumental, technical orientation, which has grown stronger in the late modern period, has fit the institutional as well as the intellectual ecology of the late modern period—not to mention its generally miserly, impatient temper. Here, modeling's *iterative* nature suits it well to the era of "release 10.7.3."

But perhaps the most important reason for modeling's continued growth and rapid diversification was that computing grew and diversified so rapidly, and the digital computer is the modeler's muse. It is, quite simply, the most perfect device yet devised for encoding, inspecting, and testing a series of models, for it operates with such speed and literalness that it can identify errors and contradictions that humans would miss. In addition, it can follow chains of operations to unthinkably distant or complex ends, often enabling the results to surprise those who programmed the model.

But the importance of digital computers for modeling goes beyond this already massive role. The personal computer spread and its software diversified (and grew more powerful and sophisticated) in a benign regress parallel to modeling's growth and diversification. This post-PC (mid-1980s and after) software explosion meant that mathematical tools of value to modelers increasingly were encoded in readily accessible software, such as now-widespread statistical analysis software packages like SPSS and Mathematica, some of which required less daunting levels of expertise to master than the original mathematics. In addition, the enormous amounts of money invested in computer hardware and software meant that by connecting social science to the world of computing, modeling opened doors not only to rigor but also to money.

Moreover, daily experience with using multiple pieces of software on one's personal computer exposed a generation to the basic concept of software as a kind of *virtual machine*—it took a certain kind of input and transformed it into a certain kind of output according to certain rules, only that input, that output, and those rules all happened to be encoded in symbols. This transposition of the idea of the machine to the realm of the virtual was a favored trope of fevered high modernists. It became an everyday part of life in the late 1980s and 1990s, when it not only appeared in countless popular films but also figured prominently in urgent legal debates about the status of "business method" (later, "business model") patents.[58] Thus, the everyday encounter with PCs running multiple pieces of software was an encounter with a particular type of manipulable mobile and an education in the powers and potential of virtual machines.[59]

Furthermore, the advent of the PC (since 1981) and World Wide Web (since 1992) has generated enormous incentives to render all models and data sets digitally to make them accessible or exchangeable online. Digitization and interconnection have made models (and the data input into them) ever more manipulable and ever more mobile, at ever higher speeds and ever greater scales. One example of this shift in possibilities is that while the various *high modern* communications revolutions enabled the rapid circulation of printed texts (and their authors) among a dispersed community of modelers across the United States and certain sites in Europe, the various *late modern* communications revolutions have enabled the near-instantaneous global circulation of texts and of a chain of associated pre-prints and electronic correspondence, the archiving of those texts (and pre-prints) in searchable online databases, and the archiving of the data sets gathered during research (and the rendering of those data sets in searchable form). The NSF, for example, now requires grant proposals to specify how the data to be gathered are to be archived and made available to other researchers: the new possibilities of digital communication have become requirements. As a result, all scientific publication is beginning to partake of the "software model" in which every "publication" is a "release" in a long iterative chain. Modeling "fits" this iterative publication "model" far better than a "literary model" of social scientific work.

The intellectual technology of modeling and the physical/symbolic technology of the digital computer have gone together so well that they have come to share each other's cultural power and resources. This opens each to criticisms primarily directed at the other, as in the sketch by British comedians Mitchell and Webb, "Kill the Poor," in which a government minister asks his young aides, as a test, whether "killing all the poor" would work as a solution to the current economic problems. After a brief protest, they dutifully tap the parameters into the model on their laptop and proclaim, "No, it won't work. The computer says so." Of course, it is the model that says so, not the computer, but the two are interchangeable, especially since neither the computer nor the model will "judge them" for having run the test, "just to see if it would work." The minister then tells them off, asking what they would have done if the run had turned out differently: the reason not to kill all the poor is that it is an unconscionably immoral act, not that the numbers do not work out.[60] For the most part, however, the elision of models and computers has redounded to the modeler's benefit: to be associated with the computer is an empowering thing.

Conclusion

It is revealing that people always talk of "building" models. They do not talk of "writing" models, or "authoring" or "composing" or "calculating" or "solving" them; they do not even really talk of "observing" models, though they do speak of "testing" them and "running" them. They certainly do not refer to "finding" or "discovering" them, or of "proving" them either. The key is that this way of speaking indicates that modelers and their audiences think of models as machines—as abstract, virtual devices with real-world importance—a conception that clearly draws on the example of computer programs as very real, virtual machines.

This way of speaking implies a vision of science that simultaneously privileges the instrumental and the abstract and that is fundamentally both conventionalist and empiricist. It is pragmatism formalized, operationalized, and assigned a payoff function. As such it may lead us to optimize our production of milk and honey, but whether it can lead us to the promised land is another question entirely.

History and Legacy, the Tree and the Web

The process of categorizing involves, if you will, an act of invention.

Jerome Bruner, Jacqueline Goodnow, and George Austin, 1956

Historians sometimes joke that biographies are really about the biographer, not the supposed subject. While this may not be wholly true, there is no doubt that all histories are products of the historian and his or her concerns every bit as much as they are descriptions of what happened. This is doubly true when the subject is recent history, and trebly true when it is recent intellectual history. The story told in these pages is thus very much the story of my own intellectual formation and that of my primary fields of study, the histories of technology and of social thought.

Because all these histories—my subjects', my fields', and my own—are intertwined, a brief excursion into historiography serves as more than just intellectual armor-plating. Rather, it may give a useful review of this book's key arguments, conclusions, and significance. In particular, it shows how putting the history of social science and the history of technology into the same frame brings a bigger picture into focus.

Since the mid-1990s there has been a surge of interest in the history of the postwar social sciences, especially of those areas where social science became part of the apparatus of state policymaking and governance.[1] These newer works have built on earlier studies, but they differ from them in both the periods they cover and the perspectives they embrace.

The history of the social sciences first became a recognizable subfield during the 1970s and 1980s, as scholars applied the questions and methods of contextualist intellectual history to the formation of the modern social sciences. Such studies combined the social history of ideas and the social history of intellectuals, focusing on the Progressive Era and interwar period.[2] This temporal focus was a logical product of their approach: to such scholars, the really interesting questions had to do with the professionalization of the social sciences and the establishment of the bounds of the modern social science disciplines, and both of these social transformations took place between roughly 1870 and 1940.[3]

A more internalist intellectual history of the changing theories and conceptual schemes of the social sciences flourished at the time as well, for the widespread perception that their fields were in crisis led many senior (or simply ambitious) social scientists to reform those fields by giving them a new history—and thus a new trajectory.[4] A few great works did both, linking intellectual transformations to professionalization and discipline formation, with probably the most important exemplars being Dorothy Ross's magisterial *The Origins of American Social Science*, Mary Furner's provocative *Advocacy and Objectivity*, Roger Bannister's *Sociology and Scientism*, and Thomas Haskell's *The Emergence of Professional Social Science*.[5]

Histories of social science written since the mid-1990s have changed direction. These studies have brought the *constructivist* perspective characteristic of the recent generation of history of science to the history of the social sciences, and with this perspective have come new questions and methods. One of the most obvious changes has been a shift in the "hot" period to study, with the post–World War II era now receiving the kind of attention that the Progressive Era once received. As a result, studies of Cold War science and society have a closer relevance to the history of the social sciences, and vice versa.[6]

This shift in temporal focus matched the shift in the questions constructivist scholars found most interesting. Their works generally focused more on interdisciplinary movements of people, practices, tools, and terms than on the processes of constructing disciplinary boundaries or of winning professional status. They looked at instruments as often as at ideas, and they paid particular attention to institutional forms (interdisciplinary research institutes, research teams, think tanks) and patronage networks (private foundations, military research agencies, and civilian research agencies) that characterized the postwar period.[7] In a nutshell, whereas the earlier generation looked at the social sciences and saw organizations, boundaries, and the creation of stable orders, the newer generation sees networks, flows, and exchanges, along with the continuous management, maintenance, and renegotiation of relationships. (In other words, the historiography of the social sciences parallels their history.)

This shift has also seen an interest in large "technosocial projects," a topic historians of Cold War science and society more generally have found fascinating as well. By *techno-social projects* I mean technical projects in which there is a deliberate effort to reconstruct social relationships by embedding them in new technological infrastructures. Such projects have existed in many times and places and on many scales. What was novel about the Cold War era was the sheer scale of such endeavors: projects from the interstate highway system to the Apollo program drew on vast state resources to bring about broad transformations of self and society. In this perspective, much of Cold War science was technoscience, and its politics was often technopolitics.[8]

Historians of technology often use the term *socio-technical systems* to refer to such hybrid projects because it is important to them to convey that the technical is also social. For other audiences, however, I prefer the term *technosocial* in order to emphasize that these projects involved the quite deliberate reconstruction of social relationships through technological means. Ideas, practices, and behaviors all were altered by new technological infrastructures—devices, systems, rules, and procedures.[9]

Such Cold War–era projects were diverse, ranging from missile defense systems to planned cities to surveillance systems to international development projects.[10] This diversity makes it difficult to generalize about them, but they did have a few things in common. First, and foremost, they all involved the mobilization of vast, diverse resources over relatively long periods of time (years, not weeks) toward the achievement of goals that were simultaneously material and social. (To adapt John Law's phrase, they were efforts at "heterogeneous social engineering," on a grand scale.)[11] In the Cold War context, these goals always were understood and legitimated in relation to the global superpower struggle, but there were a great many ways to do so in Western nations. Options for legitimation were far more limited in the Soviet bloc and the Third World.

Such projects aimed at material transformations, yes, but they also had larger goals. They were intended to produce systems, infrastructures, and institutions—things we might call patterns of practice. As a result, they posed new kinds of managerial, technical, and political challenges, involved new actors, created new relationships, and, collectively, transformed daily life in America and around the globe. As such, they were vital to the production of both the material and the political culture of the Cold War era.

In such projects, the lines between science and technology frequently were blurred, as were those between basic and applied science, and, more fundamentally, those between technologies and social institutions. It is the study of projects like these that led Langdon Winner, for example, to redefine technologies not as material objects but as "ways of building order in the world."[12] Similarly, Thomas Hughes, the dean of American historians of technology in the 1970s and 1980s, found that studying such projects required redefining technologies as systems containing both human and material-technical components. (He also found large technological systems to be crucial to modernity in general and not just to the postwar period.)[13]

Even the divide between humans and machines became porous in such projects. If an organism is but an organization and a technology simply a function given form, then it was no great leap to claim that humans and machines were species of the same genus. For instance, Dean Wooldridge, the "W" of TRW (the company that managed the Apollo moon project), wrote a series of books based on the idea that humans are simply biological machines with brains that worked much like large organizations.[14] In addition, at this same time, the concepts of structure, function, system, organization, hierarchy, communication, process, and program—the conceptual heart of the bureaucratic worldview—all were formalized in ways that made them applicable to men, machines, and institutions, formalizations that spread rapidly in the 1950s. These abstracted concepts and formalized methods quickly became common currency in the cybernetic borderland between social science, biology, and engineering, especially in the "cyborg sciences"—those postwar sciences oriented around the construction and maintenance of human-machine systems. These sciences analyzed the mechanisms of communication, command, and control in both humans and machines and so served as crucial intellectual and technical supports for the management of such large technosocial projects.[15]

Thus, the history of the social sciences in this period was intimately intertwined with that of technosocial projects, with the concepts and practices of high modern social science being both crucial supports for and characteristic products of such endeavors. These projects, in turn, were characteristic of the organizational revolution as it played out in the Cold War United States.

Historians of social science and technology in this period have found such projects, and their characteristic mingling of the technical and the social, to be both typical and unique: typical in that the technical has come to be seen as "always already" social, in all times and places; unique in the particular constellation (and scale) of technosocial ideas, institutions, and practices in this period. That is, the most striking features of postwar technosocial science have been both universalized into features of all scientific and technical pursuits and localized as products of particular people in particular places at particular points in time.

Perhaps the best example of work in this area is Paul Edwards's *The Closed World*, which brought many of these strands together. Edwards linked psychoacoustics and computing and military grand strategy and electronic surveillance and cognitive psychology and even science fiction films, following flows of people, concepts, instruments, and resources through a web of interconnected institutions, patrons, and laboratories. *The Closed World* was a hybrid book about systems science, written by a student of the author of "A Manifesto for Cyborgs."[16]

Studies like Edwards's suggest that Cold War science may be understood, at least in part, as a set of related, but distinct, efforts to reconstruct self and society through large technosocial projects.[17] The human sciences were key parts of such projects and so were important aspects of Cold War science and culture more generally, meaning that the history of the postwar social sciences has much to offer the history of Cold War science, technology, and society.

Many of these studies of postwar technosocial science are brilliant, eye-opening works. But there is something missing in them. It is not an error so much as a missing piece, without which the shape of the whole cannot be seen. That piece is an understanding of how the sciences of systems (or cyborgs) and their associated technosocial projects were related to the core mission of American social science: to assess and improve the prospects of democracy in the modern world, a task that required discovering the nature of modernity and grasping the role of reason in human affairs. Uncovering that history, sometimes hard to see amid the dollars and equations, was my goal here.

Legacy and Meaning

And what was that history? That history was the rise and eventual fracture of high modern social science, an approach to the study of humankind that redefined the central concepts, methods, tools, practices, and institutional relations of postwar social science. The exponents of this high modern style framed all subjects of study as complex, hierarchic systems defined more by their structures than by their components. They sought to construct formal models of system behavior, and they worked with eager conviction to embody those models in computer programs and simulations, embracing a conception of models as manipulable mobiles—shared tools for simplified representation and rule-governed manipulation. They developed a model of "man" as *homo adaptivus*—the limited chooser, the boundedly rational problem solver, the stress-resolving adaptation machine—a creature whose most characteristic qualities and greatest achievements *depended on* its inherent limits, particularly those to individual reason.

High modernists saw modernization as a staged developmental process, one increasingly driven by the state and involving a total change in all areas of life. Modernity, in this view, necessarily involved a shift from one social Gestalt to another and was the product of a changed understanding of the world and one's place in it—a new paradigm for life. Paradoxically, modernity thus became something that was both inevitable and chosen.

While they viewed individual rationality as profoundly limited, high modernists had grand hopes for *organized* intelligence and saw enormous potential in systems for producing rational choices, helping to redefine executives (and leaders generally) as decisionmakers (or as managers of decisionmaking processes). In keeping with this perspective on human nature and human science, and with their evangelical zeal, high modernists worked closely with the patrons of postwar social science to reform their fields and, they hoped, the world.

And they did. The ideas, practices, values, and social relations associated with high modernism were deeply intertwined with the ongoing Organizational Revolution and the unique context of Cold War America, both products and causes of a changed world.

The rise of this high modern social science, and of the bureaucratic worldview at its intellectual core, was closely linked to the organizational revolution, with the connections between this larger transformation of American society and this intellectual shift existing on three levels: on one level, the organization and scale of modern life became an assumption, a preoccupation, and a central research problem; on another, there were now great resources and rewards for work that fit well with the values, goals, and operational needs of the organizers of American business and politics—the builders and managers of the complex technosocial systems and projects that increasingly dominated the economy and polity; and on a third level, the advent and rapid spread of control technologies (especially electronic communications and information technologies) provided tangible models of the intangible structures of the world and a set of instruments for investigating, representing, and controlling it, serving as multipurpose "tools to think with."

The Cold War amplified these trends even as it directed them into new channels, providing vast resources in a hothouse atmosphere of urgent national need. The result lent almost every idea about humans or nature an ideological potential—a potential sometimes best seized upon, sometimes best denied.

As these references to the organizational revolution and the Cold War suggest, this approach to social science, the set of assumptions it encompassed, and the exemplary work it produced all mattered beyond the walls of academe: these were the ideas, ideals, and methods of those who advised policymakers in business and government, leading us into a War on Poverty at home and a war in Vietnam abroad; of those who trained elites in new schools of business and public administration; of those who wrote the basic textbooks from which a generation learned how the economy, society, polity, and even the mind worked; and who wrote the position papers, books, and magazine articles that helped set the terms of public discourse in an era of mass media, think tanks, and issue networks. These were the ideas that structured the management of the Apollo project that took us to the moon and the hub-and-spoke design of the air traffic system that takes us around this Earth. These were the ideas that defined the common sense of a large portion of the educated public and so were the ideas upon which Americans built the infrastructure of the lives we live today.

Limits, Choices, and Hopes

What does that history have to teach us as we cast our eyes from past to present to future? Most analyses of the "fracture" in American (and Euro-American) social thought and associated public discourse since the 1970s have focused either on the rise of identity politics (and the attendant politicization of the everyday) or on incipient globalization (and the attendant marketization of the everyday), with some accounts striving to link the two. In either case, the broad structure of the argument usually is that the focus (the political-cultural movement of identity politics or the econo-technical movement of incipient globalization) served to reinforce (or perhaps create) an ideological shift toward a neoliberal consensus in the realm of policy and an atomistic individualism (grounded, above all, in a near-absolute freedom of choice, with the choices usually being framed as consumer choices) in the realm of personal values and experiences.

These narratives are not wrong, and in the hands of masterful scholars like Daniel Rodgers or David Harvey they become powerful, persuasive accounts of public intellectual discourse over the past forty years.[18] Without disputing the crucial role of identity politics or incipient globalization (or of wealthy media empires) in the rise of neoliberalism and atomistic individualism, I suggest that a third narrative—a third dimension—is needed to paint a picture with the necessary depth. This narrative focuses on a changed discourse about the limits to human action in the world, a discourse with profound implications for the prospects of democracy.

In the 1950s and 1960s, American social scientists repeatedly emphasized that individual humans were severely limited in their her ability to receive and process information about the world: their reason was *bounded*, while the world was complex and interdependent. Unlike earlier social scientists, who had emphasized the emotional and the irrational as the limits to human reason, and unlike traditional, Burkean conservatives, who saw human limits in moral terms, these postwar "high modern" social scientists saw these limits as cognitive, natural, and *constructive*. Without these limits, we would be slaves to the particular and would not be able to abstract, categorize, generalize, find patterns, deduce principles—in short, to achieve what we think of as the pinnacles of intellectual achievement.

In addition, high modern social scientists were confident that our minds, however limited, were marvelously effective adaptive tools, especially when they were brought together, organized, and used in combination. Thus, amid their focus on individual human cognitive limits their faith in the power of organized intelligence to make a better world was preserved.

For example, take Paul Samuelson's *Economics: An Introductory Analysis*, first published in 1948 and now in its nineteenth edition—a textbook so popular its success changed the college textbook industry even as it taught millions of students how to understand the economic world. Samuelson, like other high modernists, was animated by both a faith in science and a concern about democracy's future. He strove to reconcile science and democracy through the substantive contributions of his science, by defining the proper role of the scientific expert in a modern society, and by attempting, through his textbook and his popular writings, to educate the citizenry and so make the people fit to govern themselves.

Samuelson made an argument on behalf of an activist government and Keynesian fiscal policy, an argument based not only on a distinction between economic and political liberties (the former had to be bounded in order for the latter to be realized in full, to Samuelson) but also on a distinctively high modern set of ideas about the structure and function of the economic system. Perhaps the most important of these ideas for his understanding of the proper role of government was that of a national economic system. Though the idea of a national economy, measured by national income and products accounts, seems only natural to us today, it was not common property in the late 1940s. Samuelson knew that this idea would be new to his audience, so he devoted the first two-thirds of his text—over 400 pages!—to developing the idea and explaining its consequences. In so doing, he made a strong case for seeing the national economy as something to be studied in itself, both because of its practical importance and because it operated according to a different set of rules than did firms or even whole industries—rules that were different because of its scale and because of the limited knowledge of the individual actors that made up the larger system.

This "macroeconomic" view of the economy, and the Keynesian policies that he saw following from it, both were based upon Samuelson's understanding of the economy as a complex organized system that tended to move toward an equilibrium state.[19] Comparing the economic system to a heating system, he described the government as a kind of thermostat or mechanical governor that set the level (of unemployment, say) at which equilibrium should be reached and that intervened, when necessary, to stabilize fluctuations about this equilibrium point.[20]

In the end, for Samuelson the federal government served a practical function in the economy that was strikingly similar to his vision of the intellectual function of macroeconomics and the social function of the expert economist: they all emerged from the systems they governed, integrating them, ordering them according to certain rules, enabling those systems to be more rational collectively than their components could be alone. It is thus no surprise that their rises have been inextricably intertwined.

Despite the power and reach of visions like Samuelson's, there were deep tensions in the high modern view: systems that produce rationality can be empowering of collective endeavor, but they also can be constraining and intrusive, especially if they impose a unitary vision of what is rational. The temptation to build the "one best system" is powerful. Also, such systems for producing "good" choices may have a hidden value set built into their definitions of what is rational—for all Samuelson's good intentions, it is hardly a value-neutral science that expressly advocates a full employment policy. In addition, the high modern approach often led to difficulty in understanding emotion and sincere faith, for they lie outside the calculus of bounded rationality, and those two are powerful constraints on the choices of both individuals and systems.

When the system produced manifest irrationality, inefficiency, and failure (as it so frequently did during the crises of the 1970s), the underlying premise that individual irrationality could be transformed into collective rationality came under heavy fire from every direction. One response was to carry the logic a step further—rationality lay not in any organization of human design, but in the environment, especially the market. So long as information (and money) could flow freely in markets, they would produce rational decisions no matter how irrational or emotional the individual person (or firm, or agency, or government) might be. In many ways, this was a return to an earlier free-market liberalism, but with a less optimistic view of human nature.

That is the neoliberal view that has dominated policymaking the past thirty-five to forty years in the United States and Britain, and the past twenty to twenty-five years in Europe. One vital aspect of this view is a somewhat fatalistic belief in the necessity of distributed decisionmaking. This belief is not a reflection of an optimistic faith in individual human abilities, however, except when politicians rephrase it for popular consumption. Rather, it is a deeply pessimistic view of collective decisionmaking as inherently, inevitably flawed. Because this view is not grounded in a faith in individual abilities but rather in a belief in the inevitable flaws in collective reason, in actual practice, it has not led to a faith in democracy but a faith in experts or "great men"; the more traditionally liberal version of this neoliberal faith being attached to expertise grounded in technical mastery of a field, and the more conservative version of it despairing of the power of education, no matter how technically sophisticated, to make people rational. Rather, to the conservative neoliberal, the ones to trust are the great individuals, as selected by competitive markets, not by experts—with the wealthy entrepreneur or financier being the paradigm cases.

Another response has been to reexamine the sources of irrationality and the notion of reason itself, whether in the form of studies of "heuristics and biases," or research into the role of affective/emotional/ideological attachments on one's ability to make sense of new information, or a return to the examination of values and other primary attachments, such as religious or ethnic or "racial" or national identities. The effects of these various aspects of the "self," to use Manuel Castells's term, generally are treated quite differently than high modern notions of the limits to reason: these are not cognitive limits, in the sense of limits in the information we know and our ability to process it quickly and efficiently.[21] Rather, the limits are emotions, values, identities—affective presences rather than absences. Sometimes these affective presences are treated as positive things, sometimes as negative ones. Crucially, they are not things that being part of an organization helps one overcome; rather, organizations generally magnify affect or substitute their own systems of values for the individual's, neither of which produces a larger rationality.

One curious aspect of this late modern shift is a distrust not only of "organized intelligence" but also of organized intervention in the world. The "multiplier effect" has been replaced by the "law of unintended consequences," which Charles Murray believed doomed all social interventions to failure.[22] We may be free to choose, but we are ever-more aware of the consequences, intended and otherwise, of our choices, and that awareness can be a mighty constraint, as anyone who has stood paralyzed in a supermarket, attempting to discern which was the more ethical box of cereal, can attest. For a political movement or ideology, such a distrust of ideas in action is, of course, deeply ironic and conflicted (though that is nothing new for political movements or ideologies), which may be why, while they serve as poles of attraction for pundits, academics, and public intellectuals, late modern visions seem to have a weaker purchase on the public imagination than ideas and policies grounded in optimism, exceptionalism, and the progressive transformation of nature and society—so long as they do not sound too utopian. What truly has been limited by our changed discourse on limits is our aspirations, our sense of what is possible when we work together.

In short, the high modern faith fractured in the 1970s, and the fault lines have yet to be mended. Social thought, especially on the level of public intellectual discourse, has shifted regarding how the limits to human knowledge and action are framed. As a result, we live in a conflicted age, where our late modern lives depend on a host of material, social, and intellectual infrastructures whose underlying logics and rationales no longer fit well with our prevailing beliefs.

The mismatch is not total: late modernism and high modernism are both modernisms. They are both attempts to make sense of a world of constant change, complex interdependence, and diverse subjectivities. Our high modern infrastructures make late modern choices possible. The Web of Networks is woven among the branches of the Tree of System. Nevertheless, the world built on the logic of structure and system and the world built on the logic of contingency and choice are different worlds. Today we are at home in neither. It is time to build anew.

Appendix: The Journal Survey

My research assistant, Sylwester Ratowt, and I conducted a survey of the flagship journals for economics, sociology, political science, anthropology, and psychology: the *American Economic Review*, *American Journal of Sociology*, *American Political Science Review*, *American Anthropologist*, and *Psychological Review*. The journals were sampled every five years, from 1925 to 1975, inclusive, with every research article in those years being included in the survey.

Information about the resulting 1,828 articles was entered into a Filemaker database, with each article described by one or more of roughly 150 keywords. Other descriptors indicated the kind of article being surveyed (a report on a survey, a literature review, a report on a lab or a field experiment, a presidential address, a theoretical synthesis, etc.), whether authors cited any sources of patronage for their work (and if so, which patrons), the authors' institutional affiliations, and whether the article employed any of various types of mathematics. After the two of us finished the initial coding of articles, I made a second pass through all the articles to assure consistency in labeling.

The assignment of keywords was a product of analysis as well as description. I did not simply tally whether a certain keyword was used but also assessed whether that keyword was significant to the article. Including every article that used the phrase "social system" or "function" as an example of the new social science, for example, would have resulted in over a hundred articles that used those terms *very* loosely. One could argue that this increase in loose usage is itself significant in that even the unreflective use of one term as opposed to another matters, so long as there is a discernible pattern to the usage. In the interest of making as strong a case as possible, however, I chose to exclude loose usage of terms unless other aspects of the article clearly indicated otherwise.

Similarly, while it was rare for an article to engage in a more formal analysis of systems or structures or to engage in model building *without* using those terms, there were a few occasions where keywords seemed appropriate despite not appearing in an article. To avoid the temptation to add keywords to make the articles fit my preconceptions, I strove to err on the side of undercounting, but there still were some instances where it was patently obvious that a keyword was appropriate despite the absence of that specific term in the text.

Using the four main keywords (*system, structure, function, model*) and twenty-one secondary keywords related to high modern social science, I divided the total universe of articles surveyed into the following concentric circles (each of which includes those within):

- The innermost circle of *Epitomes*, which includes articles tagged with three or more of the four main keywords ($N = 50$).
- A larger *Inner Core*, which includes articles tagged with two or more of the four main keywords ($N = 226$).

- A *Core*, which includes articles tagged with at least one of the four main keywords and one or more of the 21 secondary keywords (*N* = 444).
- An *Outer Core*, which includes articles tagged with at least one of the four main keywords, minus articles that used the term *model* without using any other of the main or secondary keywords (*N* = 597).
- A yet wider circle of *Affiliates*, which includes articles tagged with at least one of the main *or* secondary keywords, minus articles that use the term *model* without using any other relevant keyword (*N* = 849).
- The widest circle of articles that might be considered examples of high modern social science, the *Margins*, which includes articles tagged with at least one of the main or secondary keywords, with no subtractions (*N* = 924).
- The remainder, the set of *Outsiders*, which includes articles not tagged with any of the main or secondary keywords (*N* = 904).

Limitations and Biases of the Sample

The articles in these journals are *not* a perfect representation of the entire universe of social scientific research and publication in this period. Between 1925 and 1975, the number of journals in each field grew rapidly, with the major "second" journal in each field typically being established in the interwar period, and a host of new, more specialized, journals coming on the scene in the 1950s and 1960s. (Everyone ran out of money in the 1970s, and the price of paper shot up, so there was a lull in the creation of new journals during that decade.) In addition, the employment of social scientists on classified projects during World War II and the Cold War meant that a growing amount of research began to circulate outside of open academic journals. As many of the new journals were created to provide outlets for exactly those new kinds of work, especially rigorously mathematical work, which the flagship journals were slower to accept, and as other studies have shown military patronage to be strongly linked to the new social science, the present survey almost certainly *underestimates* the size and strength of the shift to high modern social science, perhaps by a significant degree.

In addition, this sample very likely *underestimates* the importance of actionable, instrumental knowledge in the new social science, as research published in more specialized journals or performed under contract to agencies with operational goals was more likely to have an explicit instrumental purpose. Conversely, it is possible that this sample overrepresents general theory, as it seems likely that journals whose audience is intended to be an entire discipline will prioritize articles that speak to general issues, while more specialized journals may value less universal theories. These results suggest that an additional study that compared flagship journals to "second" journals and to leading subfield journals would be of great interest.

General Findings

This survey confirms the rise of a new, high modern social science in the postwar period. Its central concepts were *system*, *structure*, and *function*, and the ideal product (not always realized in practice) was a *model* of a system's structures, functions, and relations, preferably one based on quantitative behavioral data. Articles tagged with at least one of these four keywords rose from a low of 7% of all articles in 1930 to a full 60% of all articles in 1970, a more than eight-fold increase, before declining to 52% in 1975.

It should be noted that even at its peak, high modern social science was not hegemonic. It clearly was the "mainstream" from 1955, being far more common than any single competing approach, but other approaches together constituted either a majority or a large minority at all times.

This heterogeneity raises an interesting question for the study of intellectual movements, disciplinary identities, paradigms, and so forth. From a broad perspective, it often is useful to describe periods and movements as largely co-extensive and cohesive, as when one refers to the "scientific revolution," the "quantum revolution," or the "Renaissance." In such cases, it is quite clear that both the "center" and the "bounds" of a field at some point are in a different place than they had been earlier. Whether one calls it a paradigm, conceptual scheme, way of knowing, research tradition, style, or what have you, the set of interesting and important problems, exemplary works, relevant concepts, methods, tools, institutions, and goals feels very different, and the cast of influential characters is new. Yet detailed studies of any period or movement (or paradigm, etc.) inevitably reveal wide individual variations as well as sharp contestation and debate, and the new center never is quite as new, or as central, as an Olympian view suggests.

While this study cannot resolve this dilemma, certain approaches helped me understand such movements. First, it was helpful to describe high modern social science in terms of widening circles of loosening allegiance, each of whose size and composition could and did change over time. Second, it was useful to focus on the rise to a position of *plurality* rather than hegemony for a "core set" of works: the story is the emergence of a set of works those in the field were likely to recognize as defining the mainstream, even if they themselves disagreed with important aspects of that mainstream. Third, I found it valuable to think of a movement (or paradigm, etc.) in terms of a *family* of related terms, concepts, root metaphors, practices, and social relations, rather than as a checklist of absolute requirements. Together, these approaches helped identify significant, broad movements without totalizing them.

These approaches were particularly helpful in understanding the role of patronage, for example, where there was a clear pattern of preference, without those preferences being exclusive or perfectly consistent. While there was a clear preference by patrons for work grounded in the bureaucratic worldview from 1955 to 1965, the period of greatest and most rapid change, other kinds of work continued to receive support, and the "outsider" category returned to roughly equal levels of support by 1970. Thus, there is a clear and meaningful pattern, but it is temporary and never reflects a perfect correlation.

Detailed Findings

While the survey results strongly supported my working hypotheses about the nature and trajectory of high modern social science, some of my more specific preliminary theses received ambiguous support. In particular, while works exemplifying high modern social science were *much* more likely than other pieces to explore communications systems, language, symbols, or representations (which fit my predictions nicely), works that did so were rather less common overall than I anticipated, as discussed in chapter 1. Similarly, other studies have noted a marked rise in work focusing on decisionmaking, especially on decisionmaking systems, during this period, yet very few works on choices or decisions appeared in these journals. I suspect that work on these topics (communications, symbols and representations, and decisions) appeared in other venues—such work filled virtually every issue of *Cognition*, for example, though not *Psychological Review*—but this study must leave unanswered the question as to why such work was more common in specialized journals than in the flagship journals.

In addition, the Outer Core, Core, and Inner Core's rapid increase in 1965 occurred as the percentage of works in the Affiliates category decreased (see figures 1.1–1.7); this result indicates that a loose usage of common concepts gave way to a more systematized (and even wider) usage during the heyday of high modern social science. One might have surmised incorrectly that widening use would be accompanied by looser usage. Clearly, that was not the case between 1955 and 1970, though it appears that usage began to loosen as high modern social science began to fragment in the 1970s.

The incipient decline of high modern social science in the early 1970s was heralded more by the rise of a variety of unrelated approaches and the continued shrinking of the number of loosely affiliated works than by the disappearance of works best exemplifying high modern social science. Further comparative work might be of interest here to see whether it is typical for intellectual movements to rise by bringing "friends" more firmly into the fold and to begin their decline by losing influence over their looser allies, rather than by seeing a decline in works epitomizing the core of the movement.

Patronage

Some of the most interesting results of this survey are related to patronage. First and foremost, it is very clear that patrons supported high modern social science *preferentially* but *not exclusively*. Other approaches to social science could find support, especially if they were methodologically rigorous. The most diametrically opposed approaches (moral-philosophical and public-policy-oriented) received little funding, with public-policy-oriented work receiving almost no funding whatsoever.

Although the number of articles reporting patronage is not sufficient to explain all the rise of the new approach directly, the trajectory of high modern social science matches expectations based on earlier studies of postwar patronage, with one significant exception.

In chapter 2, I argue that social science was supported by two postwar patronage systems. If the support of new postwar patrons was important to high modern social science, we should expect to see a rapid growth in high modern social science after World War II. This upturn should be first visible in the 1950 sample due to the influence of the military research agencies (especially the Office of Naval Research [ONR]), but it should be much more marked in 1955, as the Ford Foundation's program only shifted into gear after 1950. The rise should accelerate through the "golden age" of the 1960s and then taper off, possibly even declining after 1970 as the foundations wound down their programs and the federal, civilian agencies shouldered an ever-larger share of the funding burden in a time of chronic budget crises.

One also should see a shift in the 1970s away from interdisciplinary, systems-oriented research and toward other kinds of methodologically rigorous work, with rigor being defined more by internal disciplinary standards. Structural-functional, systems-oriented work and interdisciplinary work would continue, but they would be less prevalent as they no longer would be preferentially supported over other approaches.

The survey results support most of these predictions with startling accuracy. One sees a slow rise in the number of articles exemplifying high modern social science from 1925 to 1945, a modest turn upward in growth in 1950, a much more rapid rise from 1955 through 1970, and a small but clear decline in 1975. One sees a strong role for the foundations in the 1950s, a dominant role for the civilian, federal patrons from 1970 on, and a shared role in 1960 and 1965. For example, *no articles* supported by the National Science Foundation (NSF) appear in the sample before 1960, and only 5 do in that year. In 1965, 20 articles received NSF support, and 49 in 1970, however, and the latter number represented nearly a quarter of all articles published in these five journals in that year.

There also was a noticeable rise in the number of articles with multiple patrons during this golden age of 1960–70, with most of those pieces having both public and private patrons, suggesting that both public and private patrons during this period shared certain ideas about what constituted good work. In 1965, for example, 49% of all articles with patronage had two or more patrons; in 1970, 42% did. Both numbers reflect a significant rise above the earlier average of roughly 25% during 1950 and 1955. In a clear sign that high modern social science appealed to many patrons, roughly 20% of all articles in the Core had multiple patrons, while only 4% of Outsiders did. Looking at the relationship from the other direction reveals that works with multiple patrons were *extremely* likely to appear in the Core, as 54% of all articles with multiple patrons were in the Core, compared to the 22% of such articles that appeared in the Outsider category. The presence of multiple options likely meant such researchers experienced relatively high prominence in their fields, and greater independence from their patrons.

The notable divergence from predictions based on the "two systems model" is the relative absence of military patrons in the sample. Other accounts testify to the importance of the Office of Naval Research and RAND (and the Army Operations Research Office and the Air Force Office of Scientific Research and the Defense Advanced Research Projects Agency), especially during the period 1945–65, but there are *very* few articles in this sample that received military support. Only 31 articles in the entire sample of 1,828 listed any kind of military patronage, and *only 7% of the 451 articles with patronage had a military patron*, compared to 55% that had a civilian, federal patron and 51% that had a foundation patron.

Why the small military presence in this sample? While it is possible that the military role has been overstated, it seems more likely that work supported by these patrons simply appeared in other venues. For example, much work receiving direct military support is likely to have been classified or to have been published as a technical reports in the ever-growing world of "gray literature"; thus, it would have circulated via different channels from the flagship journals of academic fields: RAND published its own line of technical reports, the ONR encouraged its clients to meet with and circulate research memoranda to each other, and the SAGE air-defense project funded psychological research on human-machine interaction, much of which circulated via internal memoranda. In addition, even nonclassified work with military sponsorship was likely to have had a more applied, mission-based focus and so would be more likely to appear in a specialized journal than in a flagship journal.

Also, as suggested above, the flagship journals were slower to embrace high modern social science, especially its more mathematical forms, than newer journals were. To take but one prominent example, *Behavioral Science* began publication in 1956, and the vision statement (also published in the *American Psychologist*, another journal new to the postwar period) by the chief editor, James G. Miller, is classic expression of high modern social science. (It merits being labeled with at least 15 of my top 25 keywords!) *Econometrica* likewise was very friendly to such work during the late 1940s, 1950s, and early 1960s (though it was less friendly to the high modern style by 1970, with the exception of its embrace of modeling, which was very nearly mandatory), and it published many articles that had support from the ONR, RAND, or the Cowles Commission.

What role did the patrons of social science play in the rise of high modern social science? Did they cause its rise? Shape it? Follow it? Some insight into these questions can be gleaned by comparing the percentage of articles in the Core receiving support with the percentage of Outsider articles receiving support, over time.

Outsider articles were *more* likely to receive support in 1950 (before the Ford Foundation launched its program) than were articles in the Core. Support for articles in the Core overtook support for Outsiders in 1955 and outpaced it notably in 1960 and 1965, only to see support for Outsider pieces rise to nearly equal levels in 1970. This pattern indicates that the new patrons of social science exerted a greater influence on the practice of social science during the period 1955 to 1965/70 than before or after, as their vision of what was good social science was more distinct from previous practice.

Comparison of the connections between foundation-supported work and work supported by civilian, federal agencies is revealing. Such public patrons (the NSF and National Institutes of Health [NIH], primarily) were markedly more likely to be cited as sponsors in articles in the Core than in articles in the Outsider category (cited in 19% of Core articles versus 4% of Outsider articles). The trend for the private foundations was similar but less marked: they were cited as sponsors in nearly 9% of Core articles and just over 5% of Outsider articles. The private foundations were as likely to fund work in the Affiliates (or Margins) as in the Core or Inner Core, while the NSF and NIH were more likely to be cited in the Inner Core than in the Outer Core or Affiliates. In short, from 1965 to 1975, the NSF and NIH were *even more* closely associated with support for high modern social science than the private foundations had been at their peak period of the influence (1955 to 1965). In addition, the NSF and NIH strongly supported avowedly behavioralist work and modeling, both of which became markedly more common as those two patrons became larger players.

These facts force an adjustment to the description of the second patronage system: during the 1960s the civilian, federal patrons of social science clearly shared many commitments and assumptions with the private foundations, and they promoted them even more vigorously— and successfully. In addition, the NSF and NIH put a greater emphasis on explicit behavioralism and modeling, which fit well with the rest of high modern social science during the 1960s, though those two traits eventually diverged from their high modernist kin. Thus, the decline in high modern social science in the 1970s was influenced not only by the rise of the NSF and NIH relative to the private foundations but also by shifts *within* the NSF and NIH, as behavioralism, modeling, and methodological rigor became more highly valued even as they were less closely associated with a bureaucratic, high modern worldview.

These facts suggest two possible interpretations of the role of patronage in the rise and incipient decline of high modern social science. The first is that patronage was not a strong causal force but rather followed already existing trends in the social sciences. This possibility cannot be discounted: many of the elements of high modern social science existed before the postwar patrons came on the scene, and many leading social scientists shaped the agendas of their patrons by serving on advisory and review panels. In addition, the NSF and NIH obviously became major patrons *after* the rise of high modern social science had begun, which further supports the "patrons followed the trend" interpretation.

At the same time, the new patrons of social science chose *different* scholars to be on their boards and panels from those who had been the leaders before the war. The Ford Foundation advisory panel of 1949–50, for example, was much more a list of "who's going to be who" in the 1950s and 1960s than it was a list of "who had been who" or even "who already is who" in the 1940s. In addition, while many of the various elements of high modern social science had existed before its new patrons came on the scene, there were many possible questions and approaches in the social sciences, of which only some were chosen for support. The strong, selective support of the NSF and NIH for high modern social science made a big impact during the 1960s, helping make it the mainstream approach rather than one approach among several, and elevating the importance of explicit behavioralism and modeling within high modernist social science.

This conclusion leads us to a second possible interpretation of these results, which is that patronage influences work both directly and indirectly. Direct influence would show up clearly in this survey: patrons gave preferential support for work of a certain kind, enabling that work to be done while work that failed to receive support languished. Indirect influence, however, would not show up so clearly. Indirect influence would come as knowledge circulates of what kinds of work are favored by patrons and their advisers. Researchers then might either "chase" support (the cynical view) or internalize patrons' criteria for good science. In this way, a patron or network of patrons with clear preferences could shape the work even of researchers who never received a penny from them. Most likely, such an indirect influence would appear some short time after the direct influence, as funded work would have to circulate and be digested before it could serve as an exemplar to others.

Although such an indirect effect cannot be proven from the data in this survey, it fits many individual researchers' accounts from the period, and it may help explain why articles in the Outsider category returned to favor by 1970. Outsider articles that received support in the 1970s appear to reflect just such "chasing" of support or internalization of criteria for good science, though what they embraced were the methods (and perhaps the epistemology), not the pre-theoretical assumptions or ontology, of high modern social science.

Thus, this survey supports a picture of patronage as a tool that a certain set of social scientists used both directly and indirectly to reorient their fields toward a high modern style and to elevate explicit behavioralism and modeling within high modernism itself. This reorientation largely involved the selection and reinforcement of existing trends rather than the creation of wholly new approaches, though there were areas of genuine novelty that flowed fairly directly from the interests of patrons (work on communications; decisionmaking, game theory, or operations research; human-machine interaction; counterinsurgency; and area studies, for example). Other, more specialized, journals and classified technical reports, however, may reveal a more direct role for patrons in suggesting specific new problems for study.

Methodological Ruminations

The survey on which most of the conclusions discussed in this appendix and chapter 1 are based was a historian's survey rather than a social scientist's. I note this not to excuse myself from methodological rigor but rather to indicate that since the questions I was attempting to answer were different than a sociologist's might be, so were the goals and methods. For example, most article surveys of this size (and certainly larger ones) tend to be conducted in a somewhat mechanical fashion, with instances of use of certain words being tallied without the coder necessarily reading the actual article. I, however, did read them—enough of each one to understand its central argument and identify its core concepts and methods so that my choice of keywords would reflect their use in context rather than simply the appearance of a string of symbols. (When compared to mass keyword searches, such as Google's Ngram viewer or keyword searches of these journals in JSTOR, the results are similar on the broad level: the graph of usage over time has the same shape for "system," "structure," "process," "organization," and "behavior," along with other relevant terms, though not for "model"—there are too many articles about "supermodels" that distort the Ngram results!) These mass keyword searches were less useful for testing more specific hypotheses, however.

This survey was conducted with the goal of testing the plausibility of certain specific hypotheses rather than for the purpose of characterizing the social sciences in all their diversity. It was thus not designed to be a generative survey but an evaluative one (though reading all those articles did generate many an idea). In particular, the term list was chosen for relevance to my hypotheses rather than for relevance to other questions or as an attempt to describe the entirety of a field. And, as my hypotheses were related to the rise of a certain broad outlook over time, the terms needed to be definable in a consistent way over the course of the period *and* to be read and interpreted in context, so as to gauge whether a specific usage fit the "constant" definition or not. To do so, certain terms needed to be defined fairly broadly, with additional, specifying terms added on to deal with differentiation: hence the use of sets of related words to evaluate my basic hypotheses.

Because I attempted to trace the rise and fragmentation of a *family* of ideas and practices, certain highly specific differences in meanings and usages that matter enormously to practicing social scientists did not matter much to me. For example, my definition of *analytic realism* is pretty broad, referring to a belief that immaterial (and even unobservable) things may be valid objects of analysis provided that these "things" are defined in terms of consistent patterns of effect. Thus, to an analytic realist, gravity is an acceptable concept, even though one does not observe gravity but rather its effects. Social scientists and philosophers have argued about and elaborated on this basic idea for nearly a century now, and the different camps involved in these debates might not like being lumped together under one heading. Yet, for my purposes, what is relevant is that a family of ideas that all validated the existence and scientific relevance of "artificial," human-generated concepts, categories, or entities became widely accepted over a certain period of time. Such ideas stood in sharp contrast to Skinnerian behaviorism, for example, which denied the existence and relevance of such entities, and to older, more speculative, and less empirical social science, which accepted such concepts but did not attempt to define them in terms of mechanisms generating specific outputs from specific inputs.

Unexpected results and patterns emerged in the course of the survey, of course, leading to new questions: probably the most striking findings were (1) the relatively small number of articles that drew on communications or information theory, when other sources from the period reveal an intense interest in such topics, and (2) the great distance from application of most of the theoretical work in these articles, when other sources from the period reveal a strongly instrumental view of knowledge and a desire to use theory as a tool to intervene in the world. I interpreted these divergences from my expectations as products of the sample: the flagship journals, it seems plausible, were much more interested in general theories than in specific applications, and they were less interested in communication and information theory than more specialized journals were. A new study focusing on these ideas in practice—at the bench, in the office, in the "public sphere"—would be a vital complement to the present analysis.

Behaviorism, Behavioralism, and High Modernism

Probably the most difficult term set to deal with was the one associated with behaviorism. I wanted to distinguish between a "broad church" *behavioralism*, defined loosely as the belief that observable behavior is the only valid data source for social science, and "hard core" *behaviorism*, a belief that only observable entities have a place in science. (In other words, behavioralists hold to some form of analytic realism, while behaviorists do not.) The former is consistent with a scientific psychology that studies the mind through observable behaviors, for example, while the latter is not. Similar differences emerge in the definition of crucial, field-orienting terms like *culture, society, nation, economy,* and *market.* The latter form of behaviorism was often, though not always, associated with a Machian belief that science can establish correlations but not causes.

These two sets of terms needed to be differentiated in order to answer certain questions, but for other questions they needed to be grouped together. For example, both behaviorism- and behavioralism-related terms were indicators of adherence to a code of rigorous empirical observation, in contrast to earlier "armchair" philosophizing, and both groups tended to be allies in their pursuit of greater rigor (and more math) in social science. Behaviorists, however, were less regularly associated with the rest of the concepts and methods of high modern social science and often saw the steps toward holism in systems theories as steps back into the metaphysical past of social theory. Behavioralists, by contrast, tended to be high modernists (and vice-versa).

Another tricky part of dealing with *behavior* as a term is that the alliance of behaviorists and behavioralists largely had defeated more qualitative social science by the late 1950s to early 1960s, meaning that explicit defenses of behaviorist or behavioralist approaches largely disappeared, *except* for defenses of behavioralism versus behaviorism (and vice-versa). Over the course of the 1960s, those articles decreased in number as well, with behavioralists no longer feeling a need to defend their focus on behavior as the data, but not the subject, of social science. (Behaviorists continued to publish methodological salvos against behavioralists, but they do not appear to have disturbed their targets.) This means that just as behavioralism became a fundamental part of the mindset of most social scientists, it began to leave less direct evidence of the sort that would show up in a survey of terms and their usage. (Implicit assumptions about what constitutes good science tend not to receive specific mention in research articles, though they do get stated directly in introductory textbooks, for example.) In the end, I chose to follow the pattern I used with other terms: explicit usage or a nearly inarguable importance for the core arguments of the article merited coding with the term, while less explicit or definitive usage did not. Fortunately, the remaining qualitative or philosophical pieces did feel a need to defend themselves, giving a sense of the prevalence of the increasingly dominant, broadly behavioralist view.

Coders and Keywords

A common issue in most surveys is consistency among coders; I certainly experienced this, as in our first pass through the journals my research assistant and I often interpreted certain identifiers differently despite our meeting regularly to discuss the meanings and uses of the terms. So I made a second pass through *all* the articles, not only the ones my assistant had coded but also the ones I had coded the first time, to bring consistent usage throughout. I then tested this consistency by looking at a several dozen articles, chosen at random, for a third time. I stopped this third pass after I had made no corrections to the first three dozen reviewed.

In addition, almost all of the differences in keyword choice between myself and my assistant involved choosing a different, but closely related, term (e.g., *feedback* versus *cybernetics*, *algorithm* instead of *procedural reasoning*.) Similarly, when I checked the keywords I used to describe certain articles against the keywords used by available indexes and abstracts, the abstracts or keyword indexes usually used the same terms. When they did not, it was because they used very closely related terms (as above), usually field-specific ones I had subsumed in a broader term. Since the survey focused on the rise *not* of specific terms but rather of a certain way of thinking that was expressed in a particular family of terms (and practices), so long as my usage is consistent and so long as other reasonable observers would choose keywords from within those families of closely related terms to describe these articles, my argument should hold.

To test this belief, I substituted various combinations of terms (that seemed closely related) for specific individual terms to see if noticeably different patterns emerged. When they did, I reevaluated my definitions of the terms and paid special attention to articles coded with these terms in my second coding and analysis. For example, I began with a family of terms related to change over time; after the first pass, I realized that *evolution* was used either as a synonym for incremental development, broadly speaking, or as a specific selection process over time (and rarely both in the same article). Similarly, *process* could be used either very loosely (reflecting a general belief that change over time is ordered) or quite specifically (reflecting a direct argument that change over time is ordered by the specific mechanisms being discussed). Evolution as incremental development over time, development, and a loose usage of process all went together quite often and had very similar associations with other keywords and methods. That "development" set was distinct from those in which evolution was understood specifically as natural selection and those in which "process" had a specific mechanism attached to it (such as feedback, natural selection, or chance). Therefore, I treated these as two separate families of terms in the more general "kin-group" of discussions of change over time.

The keywords used for coding were chosen through this iterative process: based on my previous knowledge of the history of postwar behavioral science and on my hypotheses related to the emergence of a distinctive high modern social science grounded in a bureaucratic worldview, I developed an original list of about 120 terms. After going through a sample set of about 60–70 articles, we cut about a dozen terms, added about thirty more (such as *factor analysis*), and refined the definitions. After the first pass through the whole set, I made note of any problematic terms, further specified certain terms, and added a handful of new keywords to make differentiations such as the ones noted above.

The journals chosen for the survey were the flagship journals for the five largest social sciences in terms of numbers of professional members (geography could well have been included and appears to follow the same patterns). The articles surveyed within those journals were the research articles. Generally, identifying research articles was unproblematic. I did *not* include book reviews, essay reviews, bibliographies, research notes, letters to the editor, or updates on members and their activities. The only pieces that were difficult to decide whether to include were presidential addresses, which typically were not research reports but comments on the state of the field; comments and replies to comments on articles; and pieces in a few forums on major works (e.g., a forum on Rawls's *Theory of Justice* in the *APSR*). I opted to include presidential addresses because comments on the state of the field seemed relevant to this study; I included comments and replies that approximated the length of research articles but did not include ones that were markedly shorter; and I generally opted not to include articles that were parts of a forum on a particular work (as opposed to articles in a forum on a particular topic, which I did include). Because articles on major works tended to be more philosophical, this choice very slightly underrepresents philosophical approaches; however, the total number of such articles excluded is very small (less than a dozen), so the effect on the study's results is negligible.

I am happy to share my final keyword list, tables, and database on request.

Notes

Abbreviations

AA   *American Anthropologist*
AER   *American Economic Review*
AJS   *American Journal of Sociology*
APSR   *American Political Science Review*
HSP   Herbert Simon Papers, University Archives, Carnegie-Mellon University
PR   *Psychological Review*

INTRODUCTION: The Organizational Revolution and the Human Sciences

Epigraph. Myrtle McGraw, "Basic Concepts and Procedures in the Study of Behavior Development," *Psychological Review* 47, no. 1 (1940).

1. Claude Lévi-Strauss, *Structural Anthropology* (New York: Basic Books, 1963; references are to the Kindle version, locations 3367–68.) The original French edition was published in 1958.

2. The journals surveyed were the *American Economic Review, American Sociological Review, Psychological Review, American Anthropologist,* and *American Political Science Review.* The journals were sampled every five years from 1925 to 1975, inclusive, with every research article in each of these journals in those years being part of the sample, for a total of 1,828 articles. For further details on this survey, see appendix.

3. There is a large literature on market fundamentalism, choice theory, and conservative ideology. One place to begin is with Sonja Amadae, *Rationalizing Capitalist Democracy: The Cold War Origins of Rational Choice Liberalism* (Chicago: University of Chicago Press, 2003).

4. See E. E. Grant, "Scum from the Melting-Pot," *American Journal of Sociology* 30, no. 6 (1925). There is a startling shift in tone in from the 1920s to the 1930s, as talk of "racial stocks" and heritable group behavioral traits is perfectly acceptable in the 1920s in the *AJS,* even among authors who do not seem hateful (though some are), while by the mid-1930s, such language is much less common, though not entirely absent. There is a special issue of the *AJS* in 1935 (vol. 40, no. 4) devoted to explaining the proper way to think about race and race relations—it is now to be understood as a sociological phenomenon, not a biological one.

5. Daniel T. Rodgers, *Age of Fracture* (Cambridge, MA: Belknap Press of Harvard University Press, 2011).

6. Some of the key works on the organizational revolution are Robert Wiebe, *The Search for Order, 1877–1920* (New York: Hill & Wang, 1967); Louis Galambos, "Recasting the Organizational Synthesis: Structure and Process in the Twentieth and Twenty-First Centuries," *Business History Review* 79, no. 1 (2005); Alfred DuPont Chandler, *The Visible Hand: The Managerial Revolution in American Business* (Cambridge, MA: Belknap Press of Harvard University Press,

1977); *Scale and Scope: The Dynamics of Industrial Capitalism* (Cambridge, MA: Belknap Press of Harvard University Press, 1990); Louis Galambos, "The Emerging Organizational Synthesis in Modern American History," *Business History Review* 54, no. 3 (1970), and "Technology, Political Economy, and Professionalization: Central Themes of the Organizational Synthesis," *Business History Review* 57, no. 4 (1983); Brian Balogh, "Reorganizing the Organizational Synthesis: Federal-Professional Relations in Modern America," *Studies in American Political Development* 5 (Spring 1991); Kenneth E. Boulding, *The Organizational Revolution: A Study in the Ethics of Economic Organization* (Chicago: Quadrangle Books, 1968); and John Kenneth Galbraith, *The New Industrial State*, 4th ed. (Boston: Houghton Mifflin, 1985).

7. On sociology, especially Chicago sociology, see Martin Bulmer, *The Chicago School of Sociology: Institutionalization, Diversity, and the Rise of Sociological Research*, ed. Morris Janowitz, Heritage of Sociology Series (Chicago: University of Chicago Press, 1984); Robert Ezra Park et al., *The City* (Chicago: University of Chicago Press, 1925); Andrew Abbott, *Department and Discipline: Chicago Sociology at One Hundred* (Chicago: University of Chicago Press, 1999); and Mary Jo Deegan, *Jane Addams and the Men of the Chicago School, 1892–1918* (New Brunswick, NJ: Transaction, 1988). On political science, see Barry D. Karl, *Charles E. Merriam and the Study of Politics* (Chicago: University of Chicago Press, 1974); James Farr and Richard Seidelman, eds., *Discipline and History: Political Science in the United States* (Ann Arbor: University of Michigan Press, 1993); and John Gunnell, *The Descent of Political Theory: The Genealogy of an American Vocation* (Chicago: University of Chicago Press, 1993). On public administration, see Hunter Crowther-Heyck, *Herbert A. Simon: The Bounds of Reason in Modern America* (Baltimore: Johns Hopkins University Press, 2005). On social science at Chicago, see the above and Dorothy Ross, *The Origins of American Social Science* (Cambridge: Cambridge University Press, 1991). On the rise of experts, see above plus Robert L. Church, "Economists as Experts: The Rise of an Academic Profession in the United States, 1870–1920," in *The University in Society*, ed. Lawrence Stone (Princeton, NJ: Princeton University Press, 1974); Heinz Eulau, "Skill Revolution and Consultative Commonwealth," *American Political Science Review* 67, no. 1 (1973): 169–91; Thomas L. Haskell, *The Authority of Experts: Studies in History and Theory* (Bloomington: Indiana University Press, 1984); and Brian Balogh, *Chain Reaction: Expert Debate and Public Participation in American Commercial Nuclear Power, 1945–1975* (Cambridge: Cambridge University Press, 1991).

8. Roger Geiger, *Research and Relevant Knowledge: American Research Universities since World War II* (New York: Oxford University Press, 1993).

9. Ellen Condliffe Lagemann, *The Politics of Knowledge: The Carnegie Corporation, Philanthropy, and Public Policy* (Middletown, CT: Wesleyan University Press, 1989); Raymond Fosdick, *The Story of the Rockefeller Foundation* (New Brunswick, NJ: Transaction, 1989); Ford Foundation, "Report of the Study for the Ford Foundation on Policy and Program" (Detroit, MI, 1949); Roger Geiger, "American Foundations and Academic Social Science, 1945–1960," *Minerva* 26, no. 3 (1988): 315–41; Geiger, *Research and Relevant Knowledge*.

10. Stuart W. Leslie, *The Cold War and American Science: The Military-Industrial-Academic Complex at MIT and Stanford* (New York: Columbia University Press, 1993) and "Science and Politics in Cold War America" (presented to the Seminar of the History Department, Johns Hopkins University, Apr. 5, 1993); Scott G. Knowles and Stuart W. Leslie, "'Industrial Versailles': Eero Saarinen's Corporate Campuses for GM, IBM, and AT&T," *Isis* 92, no. 1 (2001): 1–33; Daniel Lee Kleinman, *Politics on the Endless Frontier: Postwar Research Policy in the United States* (Durham, NC: Duke University Press, 1995); Mark Solovey, *Shaky Foundations: The Politics-Patronage-Social Science Nexus in Cold War America*, Studies in Modern Science, Technology, and the Environment (New Brunswick, NJ: Rutgers University Press, 2013), "Riding Natural Scientists' Coattails onto the Endless Frontier: The SSRC and the Quest for Scientific Legitimacy,"

Journal of the History of the Behavioral Sciences 40, no. 4 (2004): 393–422, and "Introduction: Science and the State during the Cold War: Blurred Boundaries and a Contested Legacy," *Social Studies of Science*, no. 2 (2001): 165–70; Harvey Sapolsky, "Academic Science and the Military, the Years since World War II," in *The Sciences in the American Context*, ed. Nathan Reingold (Washington, DC: Smithsonian Institution Press, 1979), and *Science and the Navy: The History of the Office of Naval Research* (Princeton, NJ: Princeton University Press, 1990); David Hounshell, "The Cold War, Rand, and the Generation of Knowledge, 1946–1962," *Historical Studies in the Physical and Biological Sciences* 27, no. 2 (1997): 237–67; Audra J. Wolfe, *Competing with the Soviets: Science, Technology, and the State in Cold War America*, Johns Hopkins Introductory Series in the History of Science (Baltimore: Johns Hopkins University Press, 2013).

11. Ross, *Origins of American Social Science*; Thomas Haskell, *The Emergence of Professional Social Science: The ASSA and the Nineteenth Century Crisis of Authority* (Urbana: University of Illinois Press, 1977) and *Authority of Experts*; Dorothy Ross, *Modernist Impulses in the Human Sciences, 1870–1930* (Baltimore: Johns Hopkins University Press, 1994); Theodore Porter, *Trust in Numbers: The Pursuit of Objectivity in Science and Public Life* (Princeton, NJ: Princeton University Press, 1995); Theodore Porter and Dorothy Ross, eds., *The Modern Social Sciences*, vol. 7 of *The Cambridge History of Science* (New York: Cambridge University Press, 2003).

12. Timothy Mitchell, "Economists and the Economy in the Twentieth Century," in *The Politics of Method in the Human Sciences: Positivism and Its Epistemological Others*, ed. George Steinmetz (Durham, NC: Duke University Press, 2005).

13. George W. Stocking, *Race, Culture, and Evolution: Essays in the History of Anthropology* (Chicago: University of Chicago Press, 1982) and *American Anthropology, 1921–1945: Papers from the American Anthropologist* (Lincoln: University of Nebraska Press, 2002); Clyde Kluckhohn and Henry Murray, eds., *Personality in Nature, Society, and Culture* (New York: Knopf, 1948); A. L. Kroeber and Clyde Kluckhohn, *Culture: A Critical Review of Concepts and Definitions* (Cambridge, MA: Peabody Museum, 1952); W. Lloyd Warner, *Structure of American Life* (Edinburgh: University of Edinburgh Press, 1952); Robert Staughton Lynd and Helen Merrell Lynd, *Middletown: A Study in Contemporary American Culture* (New York: Harcourt, 1929) and *Middletown in Transition: A Study in Cultural Conflicts* (New York: Harcourt, 1937).

14. Hunter Crowther-Heyck, "Talcott Parsons, Science, and Social Relations at Harvard" (Department of the History of Science, Medicine, and Technology Colloquium, Johns Hopkins University, Baltimore, May 10, 1997); Charles Camic, "Introduction: Talcott Parsons before the Structure of Social Action," in *Talcott Parsons: The Early Essays*, ed. Charles Camic (Chicago: University of Chicago Press, 1991); Jeffrey C. Alexander, *Structure and Meaning: Relinking Classical Sociology* (New York: Columbia University Press, 1989); Talcott Parsons and Edward Shils, eds., *Toward a General Theory of Action* (Cambridge, MA: Harvard University Press, 1951); Talcott Parsons, *The Social System* (New York: Free Press, 1951); Robert K. Merton, *Social Theory and Social Structure: Toward the Codification of Theory and Research* (Glencoe, IL: Free Press, 1949); Bernard Barber, *Science and the Social Order* (Glencoe, IL: Free Press, 1952); Lawrence J. Henderson, *On the Social System; Selected Writings* (Chicago: University of Chicago Press, 1970).

15. Arthur Bentley, *The Process of Government: A Study of Social Pressure* (Chicago: University of Chicago Press, 1908); Charles Austin Beard, *An Economic Interpretation of the Constitution of the United States* (New York: Macmillan, 1913); Harold Lasswell, *Propaganda Technique in the World War* (London: K. Paul Trench Trubner; New York: Knopf, 1927); Harold Lasswell and Abraham Kaplan, *Power and Society: A Framework for Political Inquiry* (New Haven, CT: Yale University Press, 1950); Farr and Seidelman, *Discipline and History*; James Farr, John S. Dryzek, and Stephen T. Leonard, *Political Science in History: Research Programs and Political Traditions* (New York: Cambridge University Press, 1995); Karl Wolfgang Deutsch, *The Nerves of Government: Models of Political Communication and Control* (New York: Free Press of Glencoe, 1963); Ithiel

de Sola Pool, ed. *Contemporary Political Science: Toward Empirical Theory* (New York: McGraw-Hill, 1967); Ross, *Origins of American Social Science*.

16. A. F. Bentley, "Observable Behaviors," *Psychological Review* 47, no. 3 (1940): 230–53, 243.

17. Kerry Buckley, *Mechanical Man: John Broadus Watson and the Beginnings of Behaviorism* (New York: Guilford Press, 1989); Katherine Pandora, *Rebels within the Ranks: Psychologists' Critique of Scientific Authority and Democratic Realities in New Deal America* (New York: Cambridge University Press, 1997); James Capshew, *Psychologists on the March: Science, Practice, and Professional Identity in America, 1929–1969* (New York: Cambridge University Press, 1999); Crowther-Heyck, *Herbert A. Simon*; John M. O'Donnell, *The Origins of Behaviorism: American Psychology, 1870–1920* (New York: New York University Press, 1985); Mitchell G. Ash, *Gestalt Psychology in German Culture, 1890–1967: Holism and the Quest for Objectivity* (New York: Cambridge University Press, 1995); Wolfgang Kohler, "Gestalt Psychology Today," *American Psychologist* 14, no. 12 (1959): 727–34.

18. Evelyn Fox Keller, *Making Sense of Life: Explaining Biological Development with Models, Metaphors, and Machines* (Cambridge, MA: Harvard University Press, 2002); Lily E. Kay, *Who Wrote the Book of Life? A History of the Genetic Code* (Stanford, CA: Stanford University Press, 2000); Sharon E. Kingsland, *Modeling Nature: Episodes in the History of Population Ecology*, 2d ed. (Chicago: University of Chicago Press, 1995); Robert C. Olby, *The Path to the Double Helix: The Discovery of DNA* (New York: Dover, 1994); Pnina Abir-Am, " 'New' Trends in the History of Molecular Biology," *Historical Studies in the Physical and Biological Sciences* 26, no. 1 (1995): 167–96; Scott F. Gilbert et al., *A Conceptual History of Modern Embryology* (New York: Plenum Press, 1991); Gregg Mitman, Jane Maienschein, and Adele E. Clarke, "Crossing the Borderlands: Biology at Chicago," *Perspectives on Science: Historical, Philosophical, Social* 1, no. 3 (1993): 359–559; Ronald Rainger, Keith Benson, and Jane Maienschein, eds., *The American Development of Biology* (Philadelphia: University of Pennsylvania Press, 1988); Gregg Mitman, *The State of Nature: Ecology, Community, and American Social Thought, 1900–1950* (Chicago: University of Chicago Press, 1992) and "Defining the Organism in the Welfare State: The Politics of Individuality in American Culture, 1890–1950," in *Biology as Society, Society as Biology: Metaphors*, ed. Sabine Maasen (Dordrecht: Kluwer Academic, 1995); Soraya de Chadarevian, *Designs for Life: Molecular Biology after World War II* (New York: Cambridge University Press, 2002); Keith R. Benson, "Biology's Phoenix: Historical Perspectives on the Importance of the Organism," *American Zoologist* 29 (1989): 1067–74; Keith R. Benson, Jane Maienschein, and Ronald Rainger, *The Expansion of American Biology* (New Brunswick, NJ: Rutgers University Press, 1991); Donna Jeanne Haraway, *Crystals, Fabrics, and Fields: Metaphors of Organicism in Twentieth-Century Developmental Biology* (New Haven, CT: Yale University Press, 1976).

19. Alan J. Rocke, *Image and Reality: Kekulé, Kopp, and the Scientific Imagination*, Synthesis (Chicago: University of Chicago Press, 2010).

20. Suman Seth, *Crafting the Quantum: Arnold Sommerfeld and the Practice of Theory, 1890–1926*, Transformations (Cambridge, MA: MIT Press, 2010). Also see David Kaiser, *How the Hippies Saved Physics: Science, Counterculture, and the Quantum Revival* (New York: W. W. Norton, 2011).

21. Joel Isaac, "Theorist at Work: Talcott Parsons and the Carnegie Project on Theory, 1949–51," *Journal of the History of Ideas* 71, no. 2 (2010): 287–311.

22. For examples, see Henry J. Aaron, *Politics and the Professors: The Great Society in Perspective* (Washington, DC: Brookings Institution, 1978); Alice O'Connor, *Poverty Knowledge: Social Science, Social Policy, and the Poor in Twentieth-Century U.S. History* (Princeton, NJ: Princeton University Press, 2001); and Michael A. Bernstein, *A Perilous Progress: Economists and Public Purpose in Twentieth-Century America* (Princeton, NJ: Princeton University Press, 2001).

23. Warren Weaver, "Science and Complexity," *American Scientist* 36, no. 4 (1948): 536–44, 539.

24. Crowther-Heyck, *Herbert A. Simon*.

25. *Analytic realism* refers to an acceptance of the necessity and validity of human-generated concepts and categories for representing the world, even if those concepts and categories do not represent material things but rather patterns of relationships. For example, take the concept of a bureaucracy: one never can point at a bureaucracy; one only can point at a set of people who behave as if they were part of something we have called a bureaucracy. The typical way of demonstrating that such a set of entities was in fact an integrated system—and thus an ontologically valid category—was to measure energy or information flows among the components. Following this reasoning, many biologists at midcentury concluded that an organism was defined by its communication lines, just as political scientists concluded that a nation was defined by its patterns of communication. See, for example Karl Wolfgang Deutsch, *Nationalism and Social Communication; an Inquiry into the Foundations of Nationality* (Cambridge: Published jointly by the Technology Press of the Massachusetts Institute of Technology and Wiley, 1953).

26. *Weak holism* means the belief that while all phenomena are in principle reducible to law-governed interactions among elementary units, significant phenomena emerge at new levels in the hierarchy of complexity, phenomena that are not predictable simply on the basis of knowledge of those elemental units. In other words, to a weak holist one always can "factor down" but not always "predict up."

27. Jamie Cohen-Cole, *The Open Mind: Cold War Politics and the Sciences of Human Nature* (Chicago: University of Chicago Press, 2014).

28. While some high modernists were hard-core logical positivists, I would describe them more generally as "broad church" positivists: they saw themselves as experimentalists, admired formal analysis, aspired to objectivity, and believed in the unity of science, but they took a variety of positions on the finer points of positivism. See Peter Louis Galison, "The Americanization of Unity," *Daedalus* 127, no. 1 (1998): 45–72.

29. J. David Bolter, *Turing's Man: Western Culture in the Computer Age* (Chapel Hill: University of North Carolina Press, 1984); Otto Mayr, *Authority, Liberty, and Automatic Machinery in Early Modern Europe* (Baltimore: Johns Hopkins University Press, 1986); Crosbie Smith and Norton Wise, "Work and Waste: Political Economy and Natural Philosophy in Nineteenth-Century Britain (I) and (II)," *History of Science* 27, no. 3 (1989): 263–301 and no. 4 (1989): 391–449, and "Work and Waste: Political Economy and Natural Philosophy in Nineteenth Century Britain (III)," *History of Science* 28, no. 2 (1990): 221–61.

30. Anson Rabinbach, *The Human Motor: Energy, Fatigue, and the Origins of Modernity* (Berkeley: University of California Press, 1992); Laura Otis, *Organic Memory: History and the Body in the Late Nineteenth and Early Twentieth Centuries* (Lincoln: University of Nebraska Press, 1994); "The Metaphoric Circuit: Organic and Technological Communication in the Nineteenth Century," *Journal of the History of Ideas* 63, no. 1 (2002): 105–28; *Networking: Communicating with Bodies and Machines in the Nineteenth Century* (Ann Arbor: University of Michigan Press, 2002); and "The Other End of the Wire: Uncertainties of Organic and Telegraphic Communication," *Configurations* 9, no. 2 (2001): 181–206.

31. Paul Edwards, *The Closed World: Computers and the Politics of Discourse in Cold War America* (Cambridge, MA: MIT Press, 1996); N. Katherine Hayles, *How We Became Posthuman: Virtual Bodies in Cybernetics, Literature, and Informatics* (Chicago: University of Chicago Press, 1999) and "Flesh and Metal: Reconfiguring the Mindbody in Virtual Environments," *Configurations* 10, no. 2 (2002): 297–320; Fred Turner, *From Counterculture to Cyberculture: Stewart Brand, the Whole Earth Network, and the Rise of Digital Utopianism* (Chicago: University of Chi-

cago Press, 2006); Tim Lenoir, "Makeover: Writing the Body into the Posthuman Technoscape. Part One: Embracing the Posthuman," *Configurations* 10, no. 2 (2002): 203–20; "Makeover: Writing the Body into the Posthuman Technoscape. Part Two: Corporeal Axiomatics," *Configurations* 10, no. 3 (2002): 373–85; and "All but War Is Simulation: The Military-Entertainment Complex," *Configurations* 8, no. 3 (2000): 289–335; Manuel Castells, *The Rise of the Network Society* (Malden, MA: Blackwell, 1996).

32. Simon Schaffer, "Babbage's Intelligence: Calculating Engines and the Factory System," *Critical Inquiry* 21 (Autumn 1994): 203–27.

33. Emily Martin, *The Woman in the Body: A Cultural Analysis of Reproduction* (Boston: Beacon Press, 2001) and *Flexible Bodies: Tracking Immunity in American Culture from the Days of Polio to the Age of Aids* (Boston: Beacon Press, 1994).

34. Donald Worster, *Nature's Economy: A History of Ecological Ideas*, 2d ed. (Cambridge: Cambridge University Press, 1994); Peter J. Taylor, "Technocratic Optimism, H. T. Odum, and the Partial Transformation of the Ecological Metaphor after World War II," *Journal of the History of Biology* 21, no. 2 (1988): 213–44; Sharon E. Kingsland, *The Evolution of American Ecology, 1890–2000* (Baltimore: Johns Hopkins University Press, 2005).

35. The primary exception to this statement is the work of Michel Foucault, whose *Discipline and Punish* explores the links between the new bureaucratic institutions of state power created in late-eighteenth- and early-nineteenth-century France and the emergence of a new understanding of both the body and the self. Foucault's work is illuminating and insightful, but his almost exclusive emphasis on the micro-politics of "power-knowledge" relationships and the absence of individual human agency in his accounts make his work more a collection of valuable insights than a methodological model for my own work. Michel Foucault, *Discipline and Punish: The Birth of the Prison*, 2d Vintage ed. (New York: Vintage, 1995).

36. George Lakoff and Mark Johnson, *Metaphors We Live By* (Chicago: University of Chicago Press, 1980).

37. Lynn K. Nyhart, *Biology Takes Form: Animal Morphology and the German Universities, 1800–1900*, Science and Its Conceptual Foundations (Chicago: University of Chicago Press, 1995); Walter B. Cannon, *The Wisdom of the Body* (New York: W. W. Norton, 1939); William Coleman et al., *The Investigative Enterprise: Experimental Physiology in Nineteenth-Century Medicine* (Berkeley: University of California Press, 1988); Lawrence J. Henderson, *Pareto's General Sociology: A Physiologist's Interpretation* (Cambridge, MA: Harvard University Press, 1935); *Blood: A Study in General Physiology* (New Haven, CT: Yale University Press; London: H. Milford, Oxford University Press, 1928); Frederic L. Holmes, "Claude Bernard, the 'Milieu Intérieur,' and Regulatory Physiology," *History and Philosophy of the Life Sciences* 8, no. 1 (1986): 3–25; J. Andrew Mendelsohn, "Lives of the Cell," *Journal of the History of Biology* 36, no. 1 (2003): 1–37; Timothy Lenoir, *The Strategy of Life: Teleology and Mechanics in Nineteenth-Century German Biology* (Boston: D. Reidel, 1982); Gerald L. Geison et al., *Physiology in the American Context, 1850–1940* (Baltimore: American Physiological Society, distributed by Williams & Wilkins, 1987).

38. Wiebe, *Search for Order*. One also could call a control technology the technical infrastructure for stabilizing "heterogeneous engineering," to use John Law's terminology in his "Technology and Heterogeneous Engineering: the case of Portuguese Expansion," in *The Social Construction of Technological Systems*, ed. W. Bijker, T. P. Hughes, and T. Pinch (Cambridge, MA: MIT Press, 1987).

39. Veblen's essay "The Place of Science in Modern Civilization" in *The Place of Science in Modern Civilization and other Essays* (New York, Huebsch: 1919), 1–32, quote at 16, is an excellent early example of this fascination, for in it he emphasizes both the power of the extraordinary division of labor in modern industry and the remarkable devices and techniques that have been developed to coordinate this specialized labor.

40. Allen Newell and Herbert A. Simon, "Heuristic Problem Solving," *Journal of the Operations Research Society of America* 6, no. 1 (1958): 1–10.

41. Cohen-Cole, *Open Mind*. Rationality itself was redefined in peculiar ways during the Cold War. See Paul Erickson et al., *How Reason Almost Lost Its Mind: The Strange Career of Cold War Rationality* (Chicago, University of Chicago Press: 2013), and George Reisch, *How the Cold War Transformed Philosophy of Science: To the Icy Slopes of Logic* (New York, Cambridge University Press, 2005).

42. For a variety of approaches to the relationship of the Cold War to the social sciences, see the Focus section in *Isis* edited by David Kaiser and myself in 2010. Hunter Heyck and David Kaiser, "Introduction: New Perspectives on Science and the Cold War," *Isis* 101, no. 2 (2010): 362–66. Also see Audra Wolfe, *Competing with the Soviets: Science, Technology, and the State in Cold War America* (Baltimore: Johns Hopkins University Press, 2013).

CHAPTER 1: High Modern Social Science

Epigraph. Arthur Bentley, *Process of Government* (1908; reprint, Bloomington: Indiana University Press, 1949), 170. When first published in 1908, Bentley's book was a radical take on American politics; as such, it was noted and discussed but not embraced by the field until the interwar period, when it was taught (but still not entirely accepted) by many members of the Chicago School. Its reissue in 1949 was a signal that what was once outside the mainstream was now being cast as a herald of present trends by a postwar generation seeking to construct an intellectual ancestry. See, for example, the comment on Bentley (and Truman's *Governmental Process*) in R. E. Dowling, "Pressure Group Theory: Its Methodological Range," *APSR* 54, no. 4 (1960): 944–54.

1. The figures cited in this paragraph are drawn from my survey of the five leading social science journals for the period 1925–1975. The journals are *American Anthropologist, American Journal of Sociology, American Political Science Review, Psychological Review*, and *American Anthropologist*. For more about the methodology and goals of this survey, see appendix.

2. Bentley, "Observable Behaviors," *Psychological Review* 47, no. 3 (1940): 230–53, 243.

3. For overviews of postwar social science, see Mark Solovey and Hamilton Cravens, *Cold War Social Science: Knowledge Production, Liberal Democracy, and Human Nature* (New York: Palgrave Macmillan, 2012); and Theodore Porter and Dorothy Ross, eds., *The Modern Social Sciences*, vol. 7 of *The Cambridge History of Science* (New York: Cambridge University Press, 2003). Some exemplary studies that deal with issues of broad significance across the postwar social sciences are Jennifer S. Light, *From Warfare to Welfare: Defense Intellectuals and Urban Problems in Cold War America* (Baltimore: Johns Hopkins University Press, 2003); Sarah Elizabeth Igo, *The Averaged American: Surveys, Citizens, and the Making of a Mass Public* (Cambridge, MA: Harvard University Press, 2007); Jamie Cohen-Cole, *The Open Mind: Cold War Politics and the Sciences of Human Nature* (Chicago: University of Chicago Press, 2014); Joel Isaac, *Working Knowledge: Making the Human Sciences from Parsons to Kuhn* (Cambridge, MA: Harvard University Press, 2012); and Joel Isaac, "Introduction: The Human Sciences and Cold War America," *Journal of the History of the Behavioral Sciences* 47, no. 3 (2011): 225–31. There is a large literature, usually generated by practitioners, devoted to each discipline in the postwar period. For anthropology, the starting point is the work of George Stocking; for economics, Mark Blaug, Philip Mirowski, Mary Morgan, Judy Klein, and E. Roy Weintraub; for political science, James Farr, John Gunnell, and Richard Seidelman; for psychology, James Capshew and Ellen Herman; and for sociology, Charles Camic and Jeffrey Alexander.

4. Most accounts of the postwar behavioral science movement see economics, not political science, as the outlier. Pooley and Solovey, for example, have written insightfully on how and why economics was "marginal to the revolution." Their argument is accurate with regard

to the behavioral revolution, but behavioralism and high modernism in social science were not synonymous. Many economists shared a structural-systems view of the economy and embraced formal theory with a vengeance but still held themselves aloof from the kind of interdisciplinary ventures in behavioral science that appealed to the Ford Foundation and others discussed by Pooley and Solovey. See J. Pooley and M. Solovey, "Marginal to the Revolution: The Curious Relationship between Economics and the Behavioral Sciences Movement in Mid-Twentieth-Century America," *History of Political Economy* 42, suppl. 1 (2010): 199–233.

5. *APSR* was the slowest of the five journals to embrace high modern social science: in 1960, at least 29% of the articles the other four journals fit the criteria for articles in the Core of high modern social science (with 44% of those in *PR* doing so), while only 9% of the *APSR* articles did. Things changed rapidly in the 1960s, however, with 21% of *APSR* articles in 1965 being part of the Core, and 36% in 1970. Although *APSR* came late to the party, it did arrive.

6. On the views of the patrons, see chapter 2.

7. The results have been adjusted, downward, to account for the sharp rise in 1970–75 of articles that embraced modeling but that did not have any other sign of a connection to high modern social science. If one counts such articles, then the percentage of articles with at least one of these main keywords rises to 78% in 1970 and 77% in 1975.

8. Jeanne Haffner, *The View from Above: The Science of Social Space* (Cambridge, MA: MIT Press, 2013).

9. The term *high modern* is loosely based on James Scott and David Harvey, though without the pejorative evaluation Scott attaches to it. See James C. Scott, *Seeing Like a State: How Certain Schemes to Improve the Human Condition Have Failed* (New Haven, CT: Yale University Press, 1998).

10. The word *core* was used by Imre Lakatos (and others) to refer to that portion of a research tradition or disciplinary paradigm that the research community felt it had to protect. It thus had a special status in that community, intellectually and socially. *Core* is used here in a different sense, to refer to a set of articles in which a certain set of traits (concepts, keywords, practices) can be found. This "core set" is not defined or obtained by the same measures or procedures as a Lakatosian "core," though it might be an interesting study to compare them. See Imre Lakatos, *The Methodology of Scientific Research Programmes* (Cambridge: Cambridge University Press, 1978).

11. This strong correlation with theory is partly due to the fact that I have used modeling as one of the defining features of high modern social science, and nearly all exercises in modeling evince a strong emphasis on theory. (Not all, but nearly all.) The strong correlation of high modern social science with theory is thus an unavoidable product of the definition. (Score one for circular logic!) The correlation with theory is also very pronounced, however, if one attempts to eliminate "model" from the comparison, with articles categorized as attempts at theory building (but not modeling) being over four times as common in the Core set as in the Outsiders set. Not including the model overlap seems likely to produce an underestimate of the role of theory in high modern social science, while including the model overlap may produce a slight overestimate; hence, I use two values in table 1.1 above for this category, though the higher of the two is more accurate.

12. There were two articles that "sort of" fit under the "bio-social" umbrella in the Outer Core because they engaged in modeling—one of sex-linked spatial recognition traits, the other of inheritance of lactase deficiency, but those articles are what one might call "weak" members of both categories.

13. There is a large literature on cybernetics in postwar science and social science. Good places to start include Crowther-Heyck, *Herbert A. Simon*; Andrew Pickering, *The Cybernetic Brain: Sketches of Another Future* (Chicago: University of Chicago Press, 2010); Eden Medina,

Cybernetic Revolutionaries: Technology and Politics in Allende's Chile (Cambridge, MA: MIT Press, 2011); and Steve J. Heims, *The Cybernetics Group* (Cambridge, MA: MIT Press, 1991).

14. See Talcott Parsons, "An Analytical Approach to the Theory of Social Stratification," *American Anthropologist* 45, no. 6 (1940): 841–62. Interest in social stratification continued throughout the 1950s and 1960s, but with less attention to class, caste, and race and more discussion of a generalized "status" or of power and authority.

15. Herbert A. Simon, "The Classical Concepts of Mass and Force" (1947), HSP, Box 2, ff 55.

16. Paul Samuelson, *Foundations of Economic Analysis* (Cambridge, MA: Harvard University Press, 1947); Kenneth Elzinga, "The Eleven Principles of Economics," *Southern Economic Journal* 58, no. 4 (1992): 861–79; Hunter Crowther-Heyck, "Full Employment in a Free Society: Science and Democratic Values in Paul Samuelson's *Economics*" (presented at the Humanities and Technology Association Annual Meeting, Atlanta, April 1995); Paul A. Samuelson, *Economics: An Introductory Analysis*, 1st ed. (New York: McGraw-Hill, 1948). Reverence for Gibbs's achievement is a common refrain among mathematically skilled, American-born high modernists in the 1930s and 1940s.

17. Jean Piaget, *Structuralism* (New York: Basic Books, 1970). Note that Piaget comes in second, either to Freud or to Skinner, depending on the poll, in rankings by American psychologists of the "most influential" psychologists of the twentieth century. Since the most important structuralist works (to Americans) were written after World War II, the *interwar* bureaucratic world view cannot be said to have been influenced by continental structuralism. Similarly, while American linguists of the 1920s and 1930s certainly ought to have been familiar with Saussure's work, references to Saussure and structural linguistics are almost entirely absent in the social and behavioral sciences in America during this period. The similarities between early structuralist thought and interwar expressions of the bureaucratic worldview are more likely attributable to certain pan-Atlantic intellectual traditions of the late nineteenth century, such as post-Machian positivist philosophy and the set of fields and concerns that Anson Rabinbach and others have called "energetics."

18. In this vein, Herbert Simon wrote, "No artifact devised by man is so convenient for this kind of functional description as a digital computer. It is truly Protean, for almost the only ones of its properties that are detectable in its behavior (when it is operating properly!) are the organizational properties . . . A computer is an organization of elementary functional components in which, to a high approximation, only the function performed by those components is relevant to the behavior of the whole system." *The Sciences of the Artificial*, 3d ed. (Cambridge, MA: MIT Press, 1996), 18.

19. Solovey and Cravens, *Cold War Social Science*.

20. Allen Newell and Herbert A. Simon, "Computer Science as Empirical Enquiry: Symbols and Search," in *The Philosophy of Artificial Intelligence*, ed. Margaret Boden (New York: Oxford University Press, 1990), 105–32.

21. Herbert A. Simon, *Models of My Life* (New York: Basic Books, 1991), 193.

22. Claude E. Shannon, "A Symbolic Analysis of Relay and Switching Circuits" (Master's thesis, MIT, 1938).

23. An entire issue of the *American Anthropologist* (67, no. 5, 1965) is devoted to such models, which often are represented visually in figures that look like circuit diagrams.

24. John M. Jordan, *Machine-Age Ideology: Social Engineering and American Liberalism, 1911–1939* (Chapel Hill: University of North Carolina Press, 1994).

25. Several canonical high modernist works focused on power and influence, often looking at communication in relation to power and analyzing power in newly behavioralist terms: e.g., Harold Lasswell and Abraham Kaplan, *Power and Society* (New Haven, CT: Yale University Press, 1950); Herbert A. Simon, "Notes on the Observation and Measurement of Political Power,"

Journal of Politics 15, no. 4 (1953): 500–516; and Karl Deutsch, *The Nerves of Government* (New York: Free Press, 1963). Note that Deutsch specifically argues for redefining power in terms of "steering" and communication.

26. See Peter Galison, "The Americanization of Unity" *Daedalus* 127, no. 1 (1998): 45–72; Robert N. Proctor, *Value-Free Science? Purity and Power in Modern Knowledge* (Cambridge, MA: Harvard University Press, 1991); Rudolf Carnap and Max Black, *The Unity of Science* (London: K. Paul, Trench, Trubner, 1934); and Gerald Holton, "Ernst Mach and the Fortunes of Positivism in America," *Isis* 83, no. 1 (1992): 27–60.

27. Chester A. Barnard, *The Functions of the Executive* (1938; rev. ed., Cambridge, MA: Harvard University Press, 1968); William G. Scott, *Chester I. Barnard and the Guardians of the Managerial State* (Lawrence: University Press of Kansas, 1992); Brian R. Fry, *Mastering Public Administration: From Max Weber to Dwight Waldo*, Chatham House Series on Change in American Politics (Chatham, NJ: Chatham House, 1989); Herbert A. Simon, "Organization Theory: From Chester Barnard to the Present and Beyond," *Journal of Economic Literature*, 30, no. 3 (1992): 1503; W. Lloyd Warner and Norman H. Martin, *Industrial Man: Businessmen and Business Organizations* (New York: Harper, 1959).

28. Barnard, *Functions of the Executive*, 3, xxix.

29. Claude E. Shannon and Warren Weaver, *The Mathematical Theory of Communication* (Urbana: University of Illinois Press, 1949).

30. Warren Weaver, "Science and Complexity," *American Scientist* 36, no. 4 (1948): 536–44.

31. Warren Weaver, "Recent Contributions to the Mathematical Theory of Communication," 2. This is the introduction to Shannon and Weaver, *The Mathematical Theory of Communication*, reprinted separately. Page references are to a version found online at http://isites.harvard.edu/fs/docs/icb.topic933672.files/Weaver%20Recent%20Contributions%20to%20the%20Mathematical%20Theory%20of%20Communication.pdf.

32. Ibid., 11.

33. Ibid., 4, 8.

34. Ibid., 12.

35. Drawing found on a memo from Weaver dated Aug. 4, 1953, Warren Weaver Papers, Rockefeller Foundation Collection, Record Group 3.1, Series 915, Box 1, ff 5, pp. 2–3, Rockefeller Archives Center, Tarrytown, NY.

36. Weaver is also regarded as one of the fathers of work on the machine translation of languages.

37. Vannevar Bush, "As We May Think," *Atlantic Monthly*, July 1945, 101–8, 102.

38. J. C. R. Licklider, "Man-Computer Symbiosis," *IRE Transactions on Human Factors in Engineering* 1, no. 1 (1960): 4–11.

39. John von Neumann, "First Draft of a Report on the EDVAC" (June 30, 1945), reprinted (with typographical corrections) in *IEEE Annals of the History of Computing* 15, no. 4 (1993): 27–75.

40. Barnard, like Simon, was fascinated with W. Ross Ashby's cybernetic classic, *Design for a Brain*: "I have read [*Design for a Brain*] five times and I am certainly going to read it five more." Quoted by Kenneth Andrews in his introduction to the 1968 edition of Barnard, *Functions of the Executive*, xiii.

41. Taylor was the head of ARPA's computing program during the creation of the ARPANET and later the head of Xerox PARC during its development of the mouse, graphical user interface, and Ethernet. In other words, he presided over the creation of both the personal computer and the Internet. Taylor was introduced to the new cognitive psychology as an undergraduate and retained a strong interest in psychology throughout his career.

42. J. C. R. Licklider and Robert Taylor, "The Computer as a Communications Device," *Science and Technology* 76, no. 2 (Apr. 1968): 21–31.

CHAPTER 2: Patrons of the Revolution

Epigraph. George Katona, quoted in Social Science Research Council, Minutes of Meeting of Committee on Business Enterprise Research, Feb. 19–20, 1954, Social Science Research Council Papers, Box 6, Folder 225, p. 2, Rockefeller Archives Center, Tarrytown, NY.

1. Talcott Parsons to Dean Paul Buck, Oct. 11, 1945, Paul Buck Papers, UA III, 5.55.26, Dean (FAS), 1945–46, Box: "Social Sciences–Z," ff: "social sciences," Harvard University Archives, Cambridge, MA.

2. George Katona, quoted in Social Science Research Council, Minutes of Meeting of Committee on Business Enterprise Research, Feb. 19–20, 1954, Social Science Research Council Papers, Box 6, ff 225, p. 2, Rockefeller Archives Center, Tarrytown, NY; and Herbert A. Simon, "Some Strategic Considerations in the Construction of Social Science Models" (1951), HSP, Box 4, ff 120, p. 2. Similar sentiments are expressed throughout Katona, Simon, and the University of Michigan ISR, *1979 Founder's Symposium, Institute for Social Research* (Ann Arbor, MI: Institute for Social Research, 1980).

3. Simon, "Some Strategic Considerations," 2.

4. Behavioral and Social Science Survey Committee of the National Academy of Sciences, *The Behavioral and Social Sciences: Outlook and Needs* (Englewood Cliffs, NJ: Prentice-Hall, 1969).

5. Hunter Crowther-Heyck, *Herbert A. Simon: The Bounds of Reason in Modern America* (Baltimore: Johns Hopkins University Press, 2005); Ted Porter and Dorothy Ross, eds., *The Modern Social Sciences*, vol. 7 of *The Cambridge History of Science* (New York: Cambridge University Press, 2003); Mary Morgan, *The History of Econometric Ideas* (Cambridge: Cambridge University Press, 1990); Roger Backhouse and Philippe Fontaine, *The History of the Social Sciences since 1945* (Cambridge: Cambridge University Press, 2010).

6. James S. House et al., eds., *A Telescope on Society: Survey Research and Social Science at the University of Michigan and Beyond* (Ann Arbor: University of Michigan Press, 2004); Anne Frantilla, "Social Science in the Public Interest: A Fiftieth-Year History of the Institute for Social Research" (Ann Arbor: Bentley Historical Library, University of Michigan, 1998); Sarah Igo, *The Averaged American: Surveys, Citizens, and the Making of a Mass Public* (Cambridge, MA: Harvard University Press, 2007).

7. The term *patronage* is used in this chapter to refer to the provision of support for research: material, financial, and social. There was, and is, more to providing support for a researcher or a field than writing checks (hence, *funding* seems too narrow a term), but that does not mean the same kinds of patron-client relationships held in twentieth-century America as did with aristocratic patrons in early modern Europe.

8. For an excellent example, see Jeffrey C. Alexander, *Twenty Lectures: Sociological Theory since World War II* (New York: Columbia University Press, 1987).

9. Porter and Ross, *Modern Social Sciences*. Most of these references are to the same five pages. The Cambridge History is a superb series; thus, the relative inattention to patronage in the postwar period is all the more telling an indicator of the state of the field.

10. John Gunnell, *The Descent of Political Theory: The Genealogy of an American Vocation* (Chicago: University of Chicago Press, 1993); James Farr and Richard Seidelman, eds., *Discipline and History: Political Science in the United States* (Ann Arbor, MI: University of Michigan Press, 1993); Raymond Seidelman, *Disenchanted Realists: Political Science and the American Crisis, 1884–1984* (Albany: State University of New York Press, 1985); Alvin W. Gouldner, *The Coming Crisis of Western Sociology* (New York: Basic Books, 1970); Michael Bernstein, *A Perilous*

Progress: Economists and Public Purpose in Twentieth-Century America (Princeton, NJ: Princeton University Press, 2001); Robert M. Collins, *More: The Politics of Economic Growth in Postwar America* (New York: Oxford University Press, 2000); Dorothy Ross, "Changing Contours of the Social Science Disciplines," in Porter and Ross, *Modern Social Sciences*, 205–37; Philip Mirowski, *Machine Dreams: Economics Becomes a Cyborg Science* (Cambridge: Cambridge University Press, 2002); Ellen Herman, *The Romance of American Psychology: Political Culture in the Age of Experts* (Berkeley: University of California Press, 1995); Katherine Pandora, *Rebels Within the Ranks: Psychologists' Critique of Scientific Authority and Democratic Realities in New Deal America* (Cambridge: Cambridge University Press, 1997); Micaela Di Leonardo, *Exotics at Home: Anthropologies, Others, American Modernity* (Chicago: University of Chicago Press, 1998); Edward A. Purcell Jr., *The Crisis of Democratic Theory: Scientific Naturalism and the Problem of Value* (Lexington: University Press of Kentucky, 1973); Robert N. Proctor, *Value-Free Science? Purity and Power in Modern Knowledge* (Cambridge, MA: Harvard University Press, 1991); Howard Brick, *Daniel Bell and the Decline of Intellectual Radicalism: Social Theory and Political Reconciliation* (Madison, WI: University of Wisconsin Press, 1986); Richard Harvey Brown, *Social Science as Civic Discourse: Essays on the Invention, Legitimation, and Uses of Social Theory* (Chicago: University of Chicago Press, 1989); Seymour Bernard Sarason, *Psychology Misdirected* (New York: Free Press, 1981); Christopher Simpson, *The Science of Coercion: Communications Research and Psychological Warfare, 1945–1960* (New York: Oxford University Press, 1994); Christopher Simpson, ed., *Universities and Empire: Money and Politics in the Social Sciences during the Cold War* (New York: Free Press, 1998).

11. Mary Furner and Barry Supple, eds., *The State and Economic Knowledge: The American and British Experience* (Washington, DC: Woodrow Wilson International Center for Scholars and Cambridge University Press, 1990); Michael Lacey and Mary Furner, eds., *The State and Social Investigation in Britain and the United States* (Washington, DC: Woodrow Wilson International Center for Scholars and Cambridge University Press, 1993); Peter Wagner, Björn Wittrock, and Richard Whitley, *Discourses on Society: The Shaping of the Social Science Disciplines*, vol. 15 of *Sociology of the Sciences* (Dordrecht: Kluwer Academic, 1991); Porter and Ross, *Modern Social Sciences*.

12. Jill Morawski, ed., *The Rise of Experimentation in American Psychology* (New Haven, CT: Yale University Press, 1988); Jill Morawski, "Organizing Knowledge and Behavior at Yale's Institute of Human Relations," *Isis* 77 (1986): 219–42; Barry Karl and Stanley Katz, "The American Private Philanthropic Foundation," *Minerva* 19, no. 2 (1981): 236–70; Barry D. Karl, *Charles E. Merriam and the Study of Politics* (Chicago: University of Chicago Press, 1974); Dorothy Ross, *The Origins of American Social Science* (Cambridge: Cambridge University Press, 1991); Martin Bulmer, *The Chicago School of Sociology: Institutionalization, Diversity, and the Rise of Sociological Research*, ed. Morris Janowitz (Chicago: University of Chicago Press, 1984); Earlene Craver, "Patronage and the Directions of Research in Economics: The Rockefeller Foundation in Europe, 1924–38," *Minerva* 24, no. 2–3 (1986): 205–22; Donald Fisher, *Fundamental Development of the Social Sciences: Rockefeller Philanthropy and the United States Social Science Research Council* (Ann Arbor: University of Michigan Press, 1993); Raymond Fosdick, *The Story of the Rockefeller Foundation* (New Brunswick, NJ: Transaction, 1989).

13. Porter and Ross, *Modern Social Sciences*; Craufurd Goodwin, ed., "Economics and National Security: A History of Their Interaction," *History of Political Economy* 23, suppl (1991); Craufurd Goodwin, "The Patrons of Economics in a Time of Transformation," in *From Interwar Pluralism to Postwar Neoclassicism: Annual Supplement to Vol. 30 of History of Political Economy*, ed. Mary Morgan and Malcolm Rutherford (Durham, NC: Duke University Press, 1998): 53–81; Mark Solovey, "Project Camelot and the 1960s Epistemological Revolution: Rethinking the Politics-Patronage-Social Science Nexus," *Social Studies of Science* 31, no. 2 (2001): 171–206,

and "Riding Natural Scientists' Coattails onto the Endless Frontier: The SSRC and the Quest for Scientific Legitimacy," *Journal of the History of the Behavioral Sciences* 40, no. 4 (2004): 393–422; Roger Geiger, "American Foundations and Academic Social Science, 1945–1960," *Minerva* 26 (1988): 315–41, and *Research and Relevant Knowledge: American Research Universities since World War II* (New York: Oxford University Press, 1993); James Capshew, "Engineering Behavior: Project Pigeon, World War II, and the Conditioning of B. F. Skinner," *Technology and Culture* (1993): 835–57, and *Psychologists on the March: Science, Practice, and Professional Identity in America, 1929–1969* (New York: Cambridge University Press, 1999); Paul Edwards, *The Closed World: Computers and the Politics of Discourse in Cold War America* (Cambridge, MA: MIT Press, 1996); Peter L. Galison, "The Ontology of the Enemy: Norbert Wiener and the Cybernetic Vision," *Critical Inquiry* 21 (Autumn 1994): 228–66; Philip Mirowski, "When Games Grow Deadly Serious: The Military Influence on the Evolution of Game Theory," *History of Political Economy* 23, suppl (1991): 227–55; Peter Buck, "Adjusting to Military Life: The Social Sciences Go to War, 1941–50," in *Military Enterprise and Technological Change: Perspectives on the American Experience*, ed. Merritt Roe Smith (Cambridge, MA: MIT Press, 1985), 203–52; Mark Solovey, *Shaky Foundations: The Politics-Patronage-Social Science Nexus in Cold War America* (New Brunswick, NJ: Rutgers University Press, 2013).

14. Galison, "Ontology of the Enemy"; Andy Pickering, "Cyborg History and the World War II Regime," *Perspectives on Science* 3, no. 1 (1995): 1–48; E. Roy Weintraub, ed., *Toward a History of Game Theory*, vol. 24 supp., *Supplement to the History of Political Economy* (Durham, NC: Duke University Press, 1992); Erik Rau, "Combat Scientists: The Emergence of Operations Research in the United States During World War II" (PhD, University of Pennsylvania, 1999); M. Fortun, and Silvan Schweber, "Scientists and the Legacy of World War II: The Case of Operations Research (or);" *Social Studies of Science* 23, no. 4 (1993): 595–642; Geoffrey Bowker, "How to Be Universal: Some Cybernetic Strategies, 1943–70," *Social Studies of Science* 23, no. 1 (1993): 107–27; Andrew Abbott, *The System of Professions: An Essay on the Division of Expert Labor* (Chicago: University of Chicago Press, 1988); Stuart W. Leslie, "Playing the Education Game to Win: The Military and Interdisciplinary Research at Stanford," *Historical Studies in the Physical Sciences* 18 (1987): 55–88; Andrew Pickering, "Cybernetics and the Mangle: Ashby, Beer, and Pask," *Social Studies of Science* 32, no. 3 (2002): 413–37.

15. Samuel Klausner and Victor Lidz, eds., *The Nationalization of the Social Sciences* (Philadelphia, PA: University of Pennsylvania Press, 1986); Otto N. Larsen, *Milestones and Millstones: Social Science at the National Science Foundation, 1945–1991* (New Brunswick, NJ: Transaction, 1992); Solovey, "Riding Natural Scientists' Coattails."

16. Similar patterns can be observed in *AJS*, *AER*, and *APSR*. The changes in sociology and political science were generally somewhat slower and less complete, though they were still quite marked.

17. Paul Samuelson, "Economics in a Golden Age," in *Paul Samuelson and Modern Economic Theory*, ed. E. Cary Brown and Robert M. Solow (New York: McGraw-Hill, 1983).

18. Ross, *Origins of American Social Science*; Robert C. Bannister, *Sociology and Scientism: The American Quest for Objectivity, 1880–1940* (Chapel Hill: University of North Carolina Press, 1987); Mary Furner, *Advocacy and Objectivity: A Crisis in the Professionalization of American Social Science, 1865–1905* (Lexington: University Press of Kentucky for the OAH, 1975); Theodore Porter, *Trust in Numbers: The Pursuit of Objectivity in Science and Public Life* (Princeton, NJ: Princeton University Press, 1995).

19. Ross, "Changing Contours."

20. Crowther-Heyck, *Herbert A. Simon*.

21. Furner, *Advocacy and Objectivity*.

22. Crowther-Heyck, *Herbert A. Simon*; Louis Wirth, "Report on the History, Activities, and

Policies of the Social Science Research Council" (1937), Social Science Research Council Archives, Box 704, ff 10276, Rockefeller Archives Center.

23. Henry J. Aaron, *Politics and the Professors: The Great Society in Perspective* (Washington, DC: Brookings Institution, 1978); Frederick Mosteller and Daniel P. Moynihan, eds., *On Equality of Educational Opportunity: Papers Deriving from the Harvard University Faculty Seminar on the Coleman Report* (New York: Vintage, 1972); Daniel P. Moynihan, *Miles to Go: A Personal History of Social Policy* (Cambridge, MA: Harvard University Press, 1997); Igo, *Averaged American*; Donald T. Critchlow, *Intended Consequences: Birth Control, Abortion, and the Federal Government in Modern America* (New York: Oxford University Press, 1999); Walter Heller, *New Dimensions of Political Economy* (Cambridge, MA: Harvard University Press, 1966); Bernstein, *Perilous Progress*; Irving Louis Horowitz, ed., *The Rise and Fall of Project Camelot: Studies in the Relationship Between Science and Practical Politics* (Cambridge: MIT Press, 1967); Solovey, "Project Camelot"; Herman, *Romance of American Psychology*; Hunter Crowther-Heyck, "The Poor War in Modern Memory: Social Science and the War on Poverty" (ms., 1993).

24. Frantilla, "Social Science in the Public Interest"; House et al., *Telescope on Society*. The main difference with the ISR is that perennial large-scale survey projects and their attendant intellectual and institutional apparatus came to play the roles typically performed by disciplines elsewhere.

25. Ross, "Changing Contours."

26. Talcott Parsons, *The Social System* (New York: Free Press, 1951); David Easton, *The Political System: An Inquiry into the State of Political Science* (New York: Knopf, 1953); Paul Samuelson, *Foundations of Economic Analysis* (Cambridge, MA: Harvard University Press, 1947).

27. Gouldner, *Coming Crisis of Western Sociology*; Daniel P. Moynihan, *Maximum Feasible Misunderstanding: Community Action in the War on Poverty* (New York: Free Press, 1969).

28. The work is by Jeffrey L. Pressman and Aaron Wildavsky (Berkeley: University of California Press, 1973).

29. Ross, "Changing Contours"; Solovey, "Project Camelot" and "Riding Natural Scientists' Coattails."

30. For a classic statement of the belief in unified behavioral science, see James Grier Miller, "Toward a General Theory for the Behavioral Sciences," *American Psychologist* 10, no. 9 (1955): 513–31.

31. Thomas Parke Hughes, *Rescuing Prometheus* (New York: Pantheon, 1998); Hunter Crowther-Heyck, "George A. Miller, Language, and the Computer Metaphor of Mind," *History of Psychology* 2, no. 1 (1999): 37–64.

32. Mario Biagioli's discussion of the role of brokers in the Medici court is surprisingly relevant to the workings of this first system. In both cases, brokers were able to advance their agendas (and themselves) by connecting patrons with clients in mutually rewarding relationships. Also, links to new sources of patronage were essential in both cases to improving the status of certain sciences and their practitioners that otherwise might have languished in the university—the mathematical mixed sciences in Galileo's case, the mathematical, behavioral sciences in the present case. The chief difference was that there was no real parallel in postwar America to the status issues connected to nobility in seventeenth-century Italy. Mario Biagioli, *Galileo, Courtier* (Chicago: University of Chicago Press, 1993).

33. See Robert M. Thrall, Clyde Coombs, and Robert L. Davis, eds., *Decision Processes* (New York: Wiley, 1954); Vernon L. Smith, "Game Theory and Experimental Economics: Beginnings and Early Influences," in *Toward a History of Game Theory: Supplement to the History of Political Economy*, ed. E. Roy Weintraub (Durham, NC: Duke University Press, 1992); M. M. Flood, "Report of a Seminar on Organization Science, Rm-709" (1951), HSP, Box 28, ff: "The RAND

Corporation—Report of a Seminar on Organization Science—1951." Other attendees included Robert Bush, Clyde Coombs, William Estes, Leon Festinger, Clifford Hildreth, Samuel Karlin, Tjalling Koopmans, Jacob Marschak, Oskar Morgenstern, Roy Radner, and Lloyd Shapley.

34. Herbert A. Simon, *Models of Man: Social and Rational; Mathematical Essays on Rational Behavior in a Social Setting* (New York: Wiley, 1957), 1.

35. Robert K. Merton, *Social Theory and Social Structure: Toward the Codification of Theory and Research* (Glencoe, IL: Free Press, 1949).

36. Miller, "Toward a General Theory for the Behavioral Sciences"; Simon, *Models of Man*.

37. Two major sources for such thinking about the importance of theory in science were L. J. Henderson and A. N. Whitehead: Lawrence J. Henderson, "An Approximate Definition of a Fact," *University of California Studies in Philosophy* 14 (1932): 179–99; Alfred North Whitehead, *Science and the Modern World* (New York: Macmillan, 1925). Henderson's notion of a "conceptual scheme" is a direct ancestor of Thomas Kuhn's concept of the paradigm, minus the important concept of the exemplar. According to Henderson, before the creation of a systematic conceptual scheme, a body of knowledge is just a mass of mere data, and is not a science, just as a pre-paradigmatic body of knowledge is not a science for Kuhn. See Thomas Kuhn, *The Structure of Scientific Revolutions* (Chicago: University of Chicago Press, 1962). For further discussion of this topic, see Hunter Crowther-Heyck, "Talcott Parsons, Science, and Social Relations at Harvard" (delivered to the Johns Hopkins University Colloquium, Baltimore, May 1997) and *Herbert A. Simon*.

38. W. W. Cooper, David Rosenblatt, and Herbert Simon, "Memorandum: Research Program of the School: Project on Intra-Firm Planning" (1950), HSP, Box 62, ff: "Materials for Autobiography, 1982"; Herbert A. Simon to Bernard Berelson, Dec. 1951, HSP, Box 4, ff 119; Bernard Berelson, "The Ford Foundation Behavioral Sciences Program: Proposed Plan for the Development of the Behavioral Sciences Program—Confidential and Preliminary Draft" (1951), HSP, Box 4, ff 136.

39. Crowther-Heyck, *Herbert A. Simon*.

40. Congressional Research Service, "Research Policies for the Social and Behavioral Sciences, Science Policy Study Background Report No. 6" (Washington, DC: Task Force on Science Policy, Committee on Science and Technology, US House of Representatives, 1986).

41. Klausner and Lidz, *Nationalization of the Social Sciences*; Larsen, *Milestones and Millstones*; President's Science Advisory Committee, Behavioral Science Subpanel of the Life Science Panel, "Strengthening the Behavioral Sciences and Improving Their Use" (Washington, DC: GPO, 1962); Henry Riecken, "Underdogging: The Early Career of the Social Sciences in the NSF," in Klausner and Lidz, *Nationalization of the Social Sciences*, 209–26; Congressional Research Service, "Research Policies for the Social and Behavioral Sciences"; National Institute of Mental Health (NIMH), Division of Extramural Research Programs, Program Analysis and Evaluation Section, *Behavioral Sciences and Mental Health: An Anthology of Program Reports* (Washington, DC: GPO, 1970).

42. Larsen, *Milestones and Millstones*; Riecken, "Underdogging."

43. John T. Wilson, "Psychology and the National Science Foundation," *American Psychologist* 7, no. 9 (1952): 497–502, 500.

44. Based on survey of articles from *PR*, *ASR*, and *APSR*, between 1945 and 1975. See appendix.

45. Geiger, "American Foundations and Academic Social Science"; Francis X. Sutton, "The Ford Foundation: The Early Years," *Daedalus* 116, no. 1 (Winter 1987): 41–91.

46. Harvey Sapolsky, *Science and the Navy: The History of the Office of Naval Research* (Princeton, NJ: Princeton University Press, 1990); Capshew, *Psychologists on the March*.

47. Horowitz, *Rise and Fall of Project Camelot*; Solovey, "Project Camelot"; Stuart W. Leslie, *The Cold War and American Science: The Military-Industrial-Academic Complex at MIT and Stanford* (New York: Columbia University Press, 1993).

48. Robert H. Haveman, ed., *Poverty Policy and Poverty Research: The Great Society and the Social Sciences* (Madison: University of Wisconsin Press, 1987); Aaron, *Politics and the Professors*; Michael B. Katz, *The Undeserving Poor: From the War on Poverty to the War on Welfare* (New York: Pantheon Books, 1989); Alice O'Connor, *Poverty Knowledge: Social Science, Social Policy, and the Poor in Twentieth-Century US History* (Princeton, NJ: Princeton University Press, 2001); Harold Silver and Pamela Silver, *An Educational War on Poverty: American and British Policy-Making, 1960–1980* (Cambridge: Cambridge University Press, 1991); Crowther-Heyck, "Poor War in Modern Memory."

49. Institute for Social Research, "Development of the Scientific Capabilities of the Institute for Social Research: An Application for Support to the Division of Institutional Development, National Science Foundation" (Ann Arbor, 1970).

50. Parsons, *Social System*; Easton, *The Political System*; Karl Wolfgang Deutsch, *Nationalism and Social Communication: An Inquiry into the Foundations of Nationality* (Cambridge, MA: MIT Press; New York: Wiley, 1953). Steve J. Heims, *The Cybernetics Group* (Cambridge, MA: MIT Press, 1991); Crowther-Heyck, *Herbert A. Simon*.

51. Sonja Amadae, *Rationalizing Capitalist Democracy: The Cold War Origins of Rational Choice Liberalism* (Chicago: University of Chicago Press, 2003).

52. Crowther-Heyck, *Herbert A. Simon*.

53. The point here is not that scholars no longer study the role of communication in integrating communities. They do. Rather, such studies of the social role of communication now are one subset of a larger set of approaches to communication, many of which do not study it in terms of its role in a social system. Take, for example, the recent fascination with the idea of "memes": atomistic thought units that, in theory, compete and evolve according to the laws of Darwinian evolution. Simpson, *Science of Coercion*; Stephen W. Littlejohn and Karen A. Foss, *Theories of Human Communication*, 8th ed. (Belmont, CA: Wadsworth, 2005); Richard Dawkins, *The Selfish Gene*, 2d ed. (New York: Oxford University Press, 1990). Donald Worster has argued for a contemporaneous transformation in ecological thought from seeing ecosystems as systems to seeing them as statistical aggregates; see his *Nature's Economy*, 2d ed. (New York: Cambridge University Press, 1994).

54. Ben Orlove, "From Culture to Knowledge: Climate, Sustainable Development, and 'Traditional Environmental Knowledge,'" (presented at the University of Oklahoma, Norman, OK, Feb. 15, 2006).

55. Geiger, "American Foundations and Academic Social Science"; Sutton, "Ford Foundation"; Robert Gleeson, "The Rise of Graduate Management Education in American Universities, 1908–1970" (PhD diss., Carnegie-Mellon University, 1997); Robert Gleeson and Steven Schlossman, "George Leland Bach and the Rebirth of Graduate Management Education in the United States, 1945–1975," *Selections* 11 (1996): 8–46.

56. Sutton, "Ford Foundation."

57. David Hounshell, "The Cold War, RAND, and the Generation of Knowledge, 1946–1962," *Historical Studies in the Physical and Biological Sciences* 27, no. 2 (1997): 237–67; David Jardini, "Out of the Blue Yonder: The Rand Corporation's Diversification into Social Welfare Research, 1946–1968" (PhD diss., Carnegie-Mellon University, 1996).

58. Ford Foundation, "Report of the Study for the Ford Foundation on Policy and Program" (Detroit, MI, 1949), 14–15.

59. Miller, "Toward a General Theory for the Behavioral Sciences."

60. Berelson, "Ford Foundation Behavioral Sciences Program," 4–6, 8; Bernard Berelson,

"The Ford Foundation Behavioral Sciences Division Report, June 1953," HSP, Box 7, ff 236, pp. 12–13.

61. Berelson, "Ford Foundation Behavioral Sciences Program," 3–4.

62. Ibid., 8, 9, 30.

63. Berelson, "Ford Foundation Behavioral Sciences Division Report, June 1953."

64. Berelson, "Ford Foundation Behavioral Sciences Program," 32, 34.

65. Arnold Thackray, "Notes toward a History," *Center for Advanced Study in the Behavioral Sciences Annual Report* (1984), 59–71; Arnold Thackray, "A Site for CASBS: East or West?" *Center for Advanced Study in the Behavioral Sciences Annual Report* (1987), 63–72.

66. "Summer Research Training Institutes: A New Council Program," *SSRC Items* (1954), in HSP, Box 6, ff 225, p. 17.

67. Crowther-Heyck, *Herbert A. Simon*.

68. NIMH, *Behavioral Sciences and Mental Health*, 3.

69. Ellen Herman, "The Career of Cold War Psychology," *Radical History Review* 63 (1995): 52–85, and *Romance of American Psychology*.

70. Capshew, *Psychologists on the March*.

71. John M. Jordan, *Machine-Age Ideology: Social Engineering and American Liberalism, 1911–1939* (Chapel Hill: University of North Carolina Press, 1994).

72. On the history, logic, and politics of clinical trials, see Harry Marks, *The Progress of Experiment: Science and Therapeutic Reform in the United States, 1900–1990* (New York: Cambridge University Press, 1997).

73. Herbert A. Simon, "Application for Research Grant to NIH" (1962), HSP, Box 31, ff: "NIH Grant 1, Detailed Budget Statements 1963–64."

74. National Institute of Mental Health, Division of Extramural Research Programs, Program Analysis Section, "Mental Health Research Grant Awards, Fiscal Years 1948–63" (Washington, DC, 1964).

75. Gerald N. Grob, *From Asylum to Community: Mental Health Policy in Modern America* (Princeton, NJ: Princeton University Press, 1991), 60–67.

76. Robert Felix, "Mental Disorders as Public Health Problems," *American Journal of Psychiatry* 106 (1949): 401–6.

77. Stephen J. Cross, "Designs for Living: Lawrence K. Frank and the Progressive Legacy in American Social Science" (PhD diss., Johns Hopkins University, 1994).

78. National Institute of Mental Health, Division of Extramural Research Programs, Program Analysis and Evaluation Section, "Training Grant Program Evaluation" (Washington, DC, 1958).

79. Ibid., 15.

80. NIMH, *Behavioral Sciences and Mental Health*.

81. Grob, *From Asylum to Community*, 67.

82. Herbert A. Simon to Stanley Yolles, NIMH director, Sept. 30, 1968, HSP, Box 1, ff 11; Victoria Harden, *Inventing the NIH: Federal Biomedical Research Policy, 1887–1937* (Baltimore: Johns Hopkins University Press, 1986); Allen Newell, "Notes on Preliminary Discussion of Funding of the Systems and Communication Sciences" (Apr. 9, 1962), HSP, Box 40, ff: "Systems and Communications Sciences Program—Correspondence, Budgets, Memos, Proposals, 1961–64"; Stephen P. Strickland, *The Story of the NIH Grants Programs* (Lanham, MD: University Press of America, 1988).

83. Strickland, *Story of the NIH Grants Programs*; Jeffrey Stine, "A History of Science Policy in the United States, 1940–1985," Science Policy Study Background Report No. 1 (Washington, DC: Task Force on Science Policy, Committee on Science and Technology, House of Representatives, 1986).

84. Daryl E. Chubin and Edward J. Hackett, *Peerless Science: Peer Review and US Science Policy* (Albany: State University of New York Press, 1990).

85. US Congress, House Committee on Science Space and Technology, Subcommittee on Science, OTA Report: "Federally Funded Research, Decisions for a Decade," Hearing before the Subcommittee on Science of the Committee on Science, Space, and Technology, Mar. 20, 1991 (Washington, DC: GPO, 1991).

86. Lily Kay, *The Molecular Vision of Life: Caltech, the Rockefeller Foundation, and the Rise of the New Biology* (New York: Oxford University Press, 1993); Ronald Edmund Doel, *Solar System Astronomy in America: Communities, Patronage, and Interdisciplinary Science, 1920–1960* (New York: Cambridge University Press, 1996); Leslie, *Cold War and American Science*; Harold Orlans, *The Nonprofit Research Institute: Its Origins, Operation, Problems, and Prospects* (New York: McGraw-Hill, 1972); Hounshell, "Cold War, RAND, and the Generation of Knowledge"; Martin J. Collins, *Cold War Laboratory: RAND, the Air Force, and the American State, 1945–1950* (Washington, DC: Smithsonian Institution Press, 2002); James Digby, *Strategic Thought at RAND, 1948–63: The Ideas, Their Origins, Their Fates* (Santa Monica, CA: RAND, 1990); Julian Richard Goldstein, *RAND: The History, Operations, and Goals of a Nonprofit Corporation*, rev. ed. (Santa Monica, CA: RAND, 1961); Jardini, "Out of the Blue Yonder"; Robert J. Leonard, "Creating a Context for Game Theory," in *Toward a History of Game Theory: Supplement to the History of Political Economy*, ed. E. Roy Weintraub (Durham, NC: Duke University Press, 1992); Angela M. O'Rand, "Mathematizing Social Science in the 1950s: The Early Development and Diffusion of Game Theory," in Weintraub, *Toward a History of Game Theory*; RAND Corporation, *RAND 25th Anniversary Volume* (Santa Monica, CA: RAND, 1973); Bruce L. R. Smith, *The RAND Corporation: Case Study of a Nonprofit Advisory Corporation*, Harvard Political Studies (Cambridge, MA: Harvard University Press, 1966); Smith, "Game Theory and Experimental Economics."

87. Kent C. Redmond and Thomas M. Smith, *From Whirlwind to MITRE: The R&D Story of the SAGE Air Defense Computer*, History of Computing (Cambridge, MA: MIT Press, 2000); Hounshell, "Cold War, RAND, and the Generation of Knowledge"; Allen Barton, "Paul Lazarsfeld and Applied Social Research: Invention of the University Applied Social Research Institute," *Social Science History* 3, no. 3–4 (1979): 4–44; Igo, "America Surveyed."

88. John Gardner to Clyde Kluckhohn, "Russian Studies," July 14, 1947, UAV 759.10, Box: "Correspondence, 1947," Harvard University Archives, Cambridge, MA.

89. Ibid.

90. Leslie, *Cold War and American Science*; Rebecca Lowen, *Creating the Cold War University: The Transformation of Stanford* (Berkeley: University of California Press, 1997); Michael A. Dennis, "A Change of State: The Political Cultures of Technical Practice at the MIT Instrumentation Laboratory and the Johns Hopkins Applied Physics Laboratory, 1930–1945" (PhD diss., Johns Hopkins University, 1990); Doel, *Solar System Astronomy in America*; Peter Louis Galison, *Image and Logic: A Material Culture of Microphysics* (Chicago: University of Chicago Press, 1997).

91. A recent work that makes a strong claim for "following the money" in the history of science generally, is David Edgerton, "Time, Money, and History," *Isis* 103, no. 2 (2012): 316–27.

CHAPTER 3: The Magical Year 1956, Plus or Minus One

Epigraph. A. F. C. Wallace, "Revitalization Movements," *AA* 58, no. 2 (1956): 264–81.

1. George A. Miller, "The Magical Number Seven, Plus or Minus Two," *PR* 63, no. 2 (1956): 81–97.

2. Walt W. Rostow, "The Take-Off into Self-Sustained Growth," *Economic Journal* 66, no. 261 (1956): 25–48. Rostow elaborated on these ideas in "The Stages of Economic Growth," *Economic History Review* 12, no. 1 (1959): 1–16.

3. Herbert A. Simon, "A Behavioral Model of Rational Choice," *Quarterly Journal of Economics* 69, no. 1 (1955): 99–118, and "Rational Choice and the Structure of the Environment," *PR* 63, no. 2 (1956): 129–38; Allen Newell and Herbert A. Simon, "The Logic Theory Machine: A Complex Information Processing System," *IRE Transactions on Information Theory* 2, no. 3 (1956): 61–79; Noam Chomsky, *Syntactic Structures* (The Hague: Mouton, 1957) and "Three Models for the Description of Language," *IRE Transactions on Information Theory* 2, no. 3 (1956): 113–24; George A. Miller and Noam Chomsky, "Finite State Languages," *Information and Control* 1, no. 2 (1958): 91–112; Jerome Bruner, Jacqueline Goodnow, and George Austin, *A Study of Thinking* (New York: Wiley, 1956); W. Ross Ashby, *An Introduction to Cybernetics* (London: Chapman & Hall, 1956); Hans Selye, *The Stress of Life* (New York: McGraw-Hill, 1956); C. West Churchman, *Introduction to Operations Research* (New York: Wiley, 1957); R. Duncan Luce and Howard Raiffa, *Games and Decisions: Introduction and Critical Survey* (New York: Wiley, 1957); Talcott Parsons, *The Department and Laboratory of Social Relations, Harvard University: The First Decade, 1946–56* (Cambridge, MA: Harvard University Press, 1957); Raymond Augustine Bauer, *How the Soviet System Works: Cultural, Psychological, and Social Themes* (Cambridge, MA: Harvard University Press, 1956); Robert Dahl, "The Concept of Power," *Behavioral Science* 2, no. 3 (1957): 201–15; Anthony Downs, *An Economic Theory of Democracy* (New York: Harper & Brothers, 1957); Claude Lévi-Strauss, *Tristes Tropiques* (Paris: Plon, 1955) and *Structural Anthropology* (Paris: Plon, 1958; trans., New York: Basic Books, 1963); William H. Whyte, *The Organization Man* (New York: Simon & Schuster, 1956).

4. In the political realm, for example, Karl Wolfgang Deutsch defined the nation-organism this way in *The Nerves of Government: Models of Political Communication and Control* (New York: Free Press, 1963).

5. Abstracted and gathered over time, the trail of choices, past, present, and future, often was called a *strategy*, a concept that enabled the collapse of a process unfolding over time into a single choice-point in the present. For more on the crucial role of decisions and choices in this new model of man, see chapter 4.

6. On "satisficing," see Herbert A. Simon, *Models of Man: Social and Rational. Mathematical Essays on Rational Behavior in a Social Setting* (New York: Wiley, 1957).

7. Homeostasis is an important concept for high modernists. For its origins and initial elaboration, see Lawrence J. Henderson, *The Fitness of the Environment: An Inquiry into the Biological Significance of the Properties of Matter* (New York: Macmillan, 1913) and *Blood; a Study in General Physiology* (New Haven, CT: Yale University Press; London: H. Milford, Oxford University Press, 1928). Walter B. Cannon, *The Wisdom of the Body* (New York: W. W. Norton, 1932); Claude Bernard, Henry Copley Greene, and Lawrence Joseph Henderson, *An Introduction to the Study of Experimental Medicine* (New York: Macmillan, 1927); and Hans Selye, "Stress and the General Adaptation Syndrome," *British Medical Journal* 1, no. 4667 (1950): 1383–92.

8. Thomas Kuhn, *The Copernican Revolution* (Cambridge, MA: Harvard University Press, 1957).

9. See, for example, Clark Hull, *The Principles of Behavior: An Introduction to Behavior Theory* (New York: D. Appleton–Century, 1943), 40, and Wendell Garner, "The Contribution of Information Theory to Psychology," in *The Making of Cognitive Science: Essays in Honor of George A. Miller* (New York: Cambridge University Press, 1988), 19–35, 22. Significantly, information theory did not appear to challenge psychology's autonomy as a discipline the way that neurophysiology did. It gave psychologists a new language to use in the description and analysis of behavior without threatening to reduce psychology to electrical engineering.

10. See Garner, "Contribution of Information Theory," and George A. Miller, "What Is Information Measurement?" *American Psychologist* 8, no. 1 (1953): 3–11. For a description of the research projects in the early 1950s of several soon-to-be leaders of cognitive psychology,

including Miller, William McDill, J. C. R. Licklider, and Ulric Neisser, see J. C. R. Licklider, George A. Miller, and Jerome Wiesner, "Psychological Research Program for the Air Force Human Resources Research Laboratories, Quarterly Progress Report" (Dec. 16, 1952), HSP, Box 6, ff 213. Contrast these experimental goals with those laid out by John B. Watson in "Psychology as the Behaviorist Views It," *PR* 20, no. 2 (1913): 158–77, esp. 175.

11. George A. Miller, "George A. Miller," in *A History of Psychology in Autobiography*, ed. Gardner Lindzey (Stanford, CA: Stanford University Press, 1989). The Human Resources Research Laboratory was funded by the Air Force. The Air Force took a great interest in understanding the behavior of human-machine systems, funding MIT's Human Resources Research Laboratory, the RAND Corporation's Systems Research Laboratory, and the Human Resources Research Institute at Maxwell Air Force Base. RAND's Systems Research Laboratory conducted fascinating studies of human-computer interaction and its implications for the coordination of information for antinuclear defense in the early 1950s. For example, see R. L. Chapman and M. G. Weiner, "The History, Purpose, and Script of COGWHEEL" (June 24, 1957), RAND Corporation paper P-1105, HSP, Box 4, ff: "Herbert Simon—RAND Corporation—The History, Purpose, and Script of COGWHEEL," and Robert L. Chapman, John L. Kennedy, Allen Newell, and William C. Biel, "The Systems Research Laboratory's Air Defense Experiments" (Oct. 23, 1957), RAND Corporation paper P-1202, HSP, Box 4, ff: "Herbert Simon—RAND Corporation—The Systems Research Laboratory's Air Defense Experiments." Paul Edwards discusses some of RAND's work in this area in *The Closed World: Computers and the Politics of Discourse in Cold War America* (Cambridge, MA: MIT Press, 1996).

12. Miller, "Magical Number Seven," 81–82.

13. Ibid., 86, 81.

14. Ibid., 92.

15. Ibid., 93.

16. George A. Miller, Eugene Galanter, and Karl Pribram, *Plans and the Structure of Behavior* (New York: Henry Holt, 1960).

17. Miller, "Magical Number Seven," 93.

18. George A. Miller, interview with author, Princeton, NJ, Apr. 13, 1993.

19. Bruner, Goodnow, and Austin, *Study of Thinking*, vii.

20. Ibid., 22. For more on strategies and programs, see Crowther-Heyck, *Herbert A. Simon*.

21. For more on metaphor in science, see Mary B. Hesse, *Models and Analogies in Science* (Notre Dame, IN: University of Notre Dame Press, 1966). For a slightly later but consistent exposition of ideas about metaphor, see George Lakoff and Mark Johnson, *Metaphors We Live By* (Chicago: University of Chicago Press, 1980). For more on models and metaphors, see chapter 6.

22. There were three main elements in Chomsky's attack on the behaviorist-statistical approach to language. First, Chomsky held that, though the statistical analysis of language as word chains might describe the end product—the sentence—fairly well, it did not describe how people actually constructed sentences. People simply could not learn sentences fast enough through the building of word chains. (In a later paper, Chomsky and Miller calculated that if people actually built sentences by chaining words together, then a person would have to spend every second of every day for 100 years learning to build such chains to acquire the mastery of natural language a five-year-old possesses.) Second, Chomsky also showed that a statistical analysis of word chains was in principle unable to account for certain features of natural language (particularly a phenomenon known as symmetric embedding). Third, he observed that people tend to remember the gist of what they hear or read more accurately than they can recall its specific wording. See Chomsky, *Syntactic Structures* and "Three Models for the Description of Language," and Chomsky and Miller, "Finite-State Languages."

23. On transformational grammar, see Chomsky and Miller, "Finite-State Languages"; Chomsky, "Three Models for the Description of Language"; and George Miller, "Some Psychological Studies of Grammar," *American Psychologist* 17, no. 11 (1962): 748–62.

24. Baars, "Interview with George A. Miller," in *The Cognitive Revolution in Psychology*, ed. Bernard Baars (New York: Guilford Press, 1986).

25. According to Paul Edwards, "Miller recalls that Newell told him that Chomsky 'was developing exactly the same kind of ideas for language that he and Herb Simon were developing for theorem proving'"; *Closed World*, 229. This proof was no great surprise, as Chomsky's mathematical analysis of language was rooted in the same mathematical tradition—Russellian symbolic logic and Boolean algebra—as was Turing's Automata theory.

26. Newell and Simon, "The Logic Theory Machine: A Complex Information Processing System." The Logic Theorist was designed to prove the theorems of Russell and Whitehead's *Principia Mathematica*. This was a curiously self-referential project, as Simon's ideas about lists and their logical relations—which were fundamental to the Logic Theorist—grew out of his understanding of formal logic (especially modern set theory). This understanding, in turn, was in large part a product of his reading of Russell and Whitehead's famous text, as well as his study of the works of Rudolf Carnap.

27. This account of Simon's development of his new model of man draws heavily from Crowther-Heyck, *Herbert A. Simon*, chapter 9.

28. Simon, *Models of Man*, passim.

29. Citations of these essays will be to the reprinted versions in *Models of Man*.

30. Simon, *Models of Man*, 241.

31. Ibid., 202, emphasis in original. Though this is Simon's first published use of the term *bounded rationality*, the concept had played a prominent role in his work since the first version of *Administrative Behavior*, written in 1942. See Herbert A. Simon, "Administrative Behavior: A Study of Decision-Making Processes in Administrative Organization" (PhD diss., University of Chicago, 1945).

32. Simon, *Models of Man*, 241, 242.

33. Ibid., 245, 246.

34. Herbert to Ward Edwards, Aug. 16, 1954, HSP, Box 5, ff 203, p. 1.

35. Simon, *Models of Man*, 243, 199.

36. W. Ross Ashby, *Design for a Brain* (New York: Wiley, 1952).

37. Simon, *Models of Man*, 256.

38. There is a parallel here to the trajectory of L. J. Henderson's interests, for he moved from studying the homeostatic mechanisms of the organism to studying the "fitness of the environment" as well. See Henderson, *Fitness of the Environment*.

39. Simon, *Models of Man*, 262, 71.

40. Simon to Edwards, 1.

41. Simon, *Models of Man*, 200.

42. Crowther-Heyck, *Herbert A. Simon*.

43. Ashby, *Introduction to Cybernetics*.

44. David Mindell, *Between Human and Machine: Feedback, Control, and Computing before Cybernetics* (Baltimore: Johns Hopkins University Press, 2002).

45. Andrew Pickering discusses Ashby's ideas at length in his excellent *The Cybernetic Brain: Sketches of Another Future* (Chicago: University of Chicago Press, 2010); see also Crowther-Heyck, *Herbert A. Simon*, chapter 9. Therefore, I pass over Ashby lightly here.

46. Henderson, *Fitness of the Environment*; Cannon, *Wisdom of the Body*.

47. Selye, "Stress and the General Adaptation Syndrome," 1384.

48. Ibid.

49. Selye, *Stress of Life*, vii.

50. Herbert A. Simon, *Administrative Behavior: A Study of Decision-Making Processes in Administrative Organization* (New York: Macmillan, 1957), 102.

51. Churchman, *Introduction to Operations Research*, 4, 6.

52. Ibid., 74. Churchman here cites Arturo Rosenblueth, Norbert Wiener, and Julian Bigelow, "Behavior, Purpose, and Teleology," *Philosophy of Science* 10 (1943): 18–24.

53. Robert Dahl, *A Preface to Democratic Theory* (Chicago: University of Chicago Press, 1956), 132, 143.

54. Luce and Raiffa, *Games and Decisions: Introduction and Critical Survey*, 13. For more on this topic, see chapter 4.

55. Anthony Downs, "An Economic Theory of Political Action in a Democracy," *Journal of Political Economy* 65, no. 2 (1957): 135–50, 135, 136.

56. Talcott Parsons and Neil Smelser, *Economy and Society* (New York: Free Press, 1956).

57. James G. March, Herbert A. Simon, and Harold Guetzkow, *Organizations* (New York: Wiley, 1958).

58. Wallace, "Revitalization Movements," *AA*, 58, no. 2 (1956): 264–81, 280.

59. Anthony F. C. Wallace, "Individual Differences and Cultural Uniformities," *American Sociological Review* 17, no. 6 (1952): 747–50.

60. Kuhn, *Copernican Revolution*, 37.

61. Ibid.

62. Ibid., 39, 76.

63. Ibid., 76.

64. Walt Rostow, "The Take-Off into Self-Sustained Growth," *Economic Journal* 66, no. 261 (1956): 25–48.

CHAPTER 4: Producing Reason

Epigraph. Herbert A. Simon to Ward Edwards, Aug. 16, 1954, HSP, Box 5, ff 203, p. 1.

1. Ward Edwards, "The Theory of Decision Making," *Psychological Bulletin* 51, no. 4 (1954): 380–417.

2. John Von Neumann and Oskar Morgenstern, *The Theory of Games and Economic Behavior* (Princeton, NJ: Princeton University Press, 1944).

3. On the idea of the "finite problem-solver" as the new model human, see Hunter Crowther-Heyck, *Herbert A. Simon: The Bounds of Reason in Modern America* (Baltimore: Johns Hopkins University Press, 2005) and "Mind and Network," *Annals of the History of Computing* 27, no. 3 (2005): 103–4.

4. For a detailed discussion of the "sciences of choice" and the "sciences of control," see Crowther-Heyck, *Herbert A. Simon*. On the sciences of choice, see Sonja Amadae, *Rationalizing Capitalist Democracy: The Cold War Origins of Rational Choice Liberalism* (Chicago: University of Chicago Press, 2003); Stephen P Waring, "Cold Calculus: The Cold War and Operations Research," *Radical History Review* 63 (Fall 1995): 28–51; Philip Mirowski, *Machine Dreams: Economics Becomes a Cyborg Science* (Cambridge: Cambridge University Press, 2002); and E. Roy Weintraub, *How Economics Became a Mathematical Science* (Durham, NC: Duke University Press, 2002).

The literature on the sciences of choice usually has been quite distinct from work on the family of fields related to cybernetics, though cybernetics and systems analysis addressed many of the same issues. See Roberto Cordeschi, *The Discovery of the Artificial: Behavior, Mind, and Machines Before and Beyond Cybernetics* (Dordrecht: Kluwer Academic, 2002); Ronald Kline, "Cybernetics, Management Science, and Technology Policy: The Emergence of 'Information Technology' as a Keyword, 1948–1985," *Technology and Culture* 47, no. 3 (2006): 513–35; N. Katherine Hayles, *How We Became Posthuman: Virtual Bodies in Cybernetics, Literature, and*

Informatics (Chicago: University of Chicago Press, 1999); Evelyn Fox Keller, "Organisms, Machines, and Thunderstorms: A History of Self-Organization, Part One," *Historical Studies in the Natural Sciences* 38, no. 1 (2008): 45–75; David Mindell, *Between Human and Machine: Feedback, Control, and Computing Before Cybernetics* (Baltimore: Johns Hopkins University Press, 2002); Andrew Pickering, "Cybernetics and the Mangle: Ashby, Beer, and Pask," *Social Studies of Science* 32, no. 3 (2002): 413–37.

5. Herbert A. Simon, *Administrative Behavior*, 2d ed. (New York: Macmillan, 1961), 102.

6. For an insightful overview of "embedded liberalism" versus "neoliberalism" that discusses choice, reason, and individual freedom, see David Harvey, *A Brief History of Neoliberalism* (Oxford: Oxford University Press, 2005).

7. Kenneth Arrow surveys several of these redefinitions and their potential uses in "Utilities, Attitudes, Choices: A Review Note," *Econometrica* 26, no. 10 (1958): 1–23.

8. Edward Jones-Imhotep, "Maintaining Humans," in Solovey and Cravens, Cold War Social Science: Knowledge Production, Liberal Democracy, and Human Nature" (New York: Palgrave Macmillan, 2012), 175–96.

9. Paul Erickson et al., *How Reason Almost Lost Its Mind: The Strange Career of Cold War Rationality* (Chicago: University of Chicago Press, 2013), 3, 2.

10. Ibid., 5.

11. This third context of longstanding debates within the social sciences about certain central issues, as reformulated in a new technical language, is a theme in David Engerman, "Social Science in the Cold War" *Isis* 101, no. 2 (June 2010): 393–400. Another valuable perspective on the historiography of recent social science is Joel Isaac, "Tangled Loops: Theory, History, and the Human Sciences in Modern America," *Modern Intellectual History* 6, no. 2 (2009): 397–424.

12. On "technopolitics," see Gabrielle Hecht, *The Radiance of France: Nuclear Power and National Identity after World War II* (Cambridge, MA: MIT Press, 1998). On "technoscience," see Don Ihde and Evan Selinger, eds., *Chasing Technoscience: Matrix for Materiality* (Bloomington: Indiana University Press, 2003); Thomas Parke Hughes, Agatha Chipley Hughes, Michael Thad Allen, and Gabrielle Hecht, eds., *Technologies of Power: Essays in Honor of Thomas Parke Hughes and Agatha Chipley Hughes* (Cambridge, MA: MIT Press, 2001); and John Law, *Aircraft Stories: Decentering the Object in Technoscience* (Durham, NC: Duke University Press, 2002). On infrastructural projects as technosocial projects, see Paul N. Edwards, "Meteorology as Infrastructural Globalism," *Osiris* 21, no. 1 (2006): 229–50, and Thomas J. Misa, Philip Brey, and Andrew Feenberg, *Modernity and Technology* (Cambridge, MA: MIT Press, 2003).

13. This argument is at the center of H. Stuart Hughes, *Consciousness and Society: The Reorientation of European Social Thought, 1890–1930* (New York: Vintage, 1958). Also see Dorothy Ross, *Modernist Impulses in the Human Sciences, 1870–1930* (Baltimore: Johns Hopkins University Press, 1994), introduction, and Edward A. Purcell Jr., *The Crisis of Democratic Theory: Scientific Naturalism and the Problem of Value* (Lexington: University Press of Kentucky, 1973).

14. Hughes, *Consciousness and Society*. Some exemplary works include Sigmund Freud, *Civilization and Its Discontents*, trans. James Strachey (New York: W.W. Norton, 1961), and Vilfredo Pareto, *Mind and Society* (New York: Harcourt, Brace, 1935).

15. Robert E. Park, Ernest W. Burgess, and Roderick D. McKenzie, eds., *The City* (Chicago: University of Chicago Press, 1967); William I. Thomas and Florian Znaniecki, *The Polish Peasant in Europe and America: A Monograph of an Immigrant Group* (Boston: R. G. Badger, 1918–1920); William Ogburn, *Social Change with Respect to Culture and Original Nature* (London: G. Allen & Unwin, 1923).

16. For Merriam and Lasswell's interest in psychology, see Barry Karl, *Charles E. Merriam and the Study of Politics* (Chicago: University of Chicago Press, 1974), 106–7, 171. Also see Harold

Lasswell, *Psychopathology and Politics* (Chicago: University of Chicago Press, 1930); Charles Merriam and Harold Gosnell, *Non-Voting, Causes and Methods of Control* (Chicago: University of Chicago Press, 1924).

17. Walter Lippmann, *Public Opinion* (New York: Harcourt, Brace, 1922); Walter Lippmann, *Drift and Mastery: An Attempt to Diagnose the Current Unrest* (New York: Kennerly, 1914).

18. Purcell, *Crisis of Democratic Theory*, esp. chapters 2–3, 10. For an insightful analysis of concerns about the rationality of the public in the age of mass communications, see Richard Butsch, *The Citizen Audience: Crowds, Publics, and Individuals* (New York: Routledge, 2008), and Jamie Cohen-Cole, *The Open Mind: Cold War Politics and the Sciences of Human Nature* (Chicago: University of Chicago Press, 2014).

19. Jacob Marschak, "Rational Behavior, Uncertain Prospects, and Measurable Utility," *Econometrica* 18, no. 2 (1950): 111–41.

20. Examples of such systems thinking include David Easton, *The Political System: An Inquiry into the State of Political Science* (New York: Knopf, 1953); Talcott Parsons, *The Social System* (New York: Free Press, 1951); and Simon, *Administrative Behavior*.

21. James Grier Miller, "Toward a General Theory for the Behavioral Sciences," *American Psychologist* 10, no. 9 (1955): 513–31.

22. Karl Wolfgang Deutsch, *The Nerves of Government: Models of Political Communication and Control* (New York: Free Press of Glencoe, 1963).

23. Warren Weaver and Claude E Shannon, *The Mathematical Theory of Communication* (Urbana: University of Illinois Press, 1949).

24. On this view of language, see Hunter Crowther-Heyck, "George A. Miller, Language, and the Computer Metaphor of Mind," *History of Psychology* 2, no. 1 (1999): 37–64.

25. James G March, Herbert A Simon, and Harold Guetzkow, *Organizations* (New York: Wiley, 1958); Simon, *Administrative Behavior*; Herbert A Simon, Donald Smithburg, and Victor Thompson, *Public Administration*, 1st ed. (New York: Knopf, 1950).

26. Peter F. Drucker, *Concept of the Corporation* (New York: John Day, 1946).

27. Robert A. Dahl and Charles Lindblom, *Politics, Economics, and Welfare* (New York: Harper & Brothers, 1953); Paul Felix Lazarsfeld, Bernard Berelson, and Hazel Gaudet, *The People's Choice: How the Voter Makes up His Mind in a Presidential Campaign*, 2d ed. (New York: Columbia University Press, 1948); Bernard Berelson, *Voting: A Study of Opinion Formation in a Presidential Campaign* (Chicago: University of Chicago Press, 1954); Sarah Elizabeth Igo, *The Averaged American: Surveys, Citizens, and the Making of a Mass Public* (Cambridge, MA: Harvard University Press, 2007).

28. Herbert A Simon, "Notes on the Observation and Measurement of Political Power," *Journal of Politics* 15, no. 4 (Nov. 1953): 500–516; Harold Lasswell and Abraham Kaplan, *Power and Society: A Framework for Political Inquiry* (New Haven, CT: Yale University Press, 1950).

29. Kenneth Joseph Arrow, *Social Choice and Individual Values* (New York: Wiley, 1951); P. A. Samuelson, "Consumption Theory in Terms of Revealed Preference," *Economica* 15, no. 60 (1948): 243–53, and "Social Indifference Curves," *Quarterly Journal of Economics* 70, no. 1 (1956): 1–22.

30. Herbert A. Simon, "A Comparison of Game Theory and Learning Theory," *Psychometrika* 21, no. 3 (Sept. 1956): 267–72, and "Theories of Decision-Making in Economics and Behavioral Science," *American Economic Review* 49, no. 3 (1959): 253–83.

31. James March, "An Introduction to the Theory and Measurement of Influence," *APSR* 49, no. 2 (1955): 431–51, 431.

32. R. Duncan Luce and Howard Raiffa, *Games and Decisions: Introduction and Critical Survey* (New York: Wiley, 1957).

33. Abraham Wald, *Statistical Decision Functions* (New York: Wiley, 1950).

34. Arrow, "Utilities, Attitudes, Choices," 2.

35. Here, *social reform* refers to the attempt to solve widely recognized "social problems" (such as unemployment or delinquency or race relations) through the adaptation and improvement of existing institutions rather than through radical social change. To the cohort of social scientists being discussed in this paper, social reform usually was to be achieved by better planning, better organization, and better management.

36. Irving Louis Horowitz, ed. *The Rise and Fall of Project Camelot: Studies in the Relationship between Science and Practical Politics* (Cambridge: MIT Press, 1967); Mark Solovey, "Project Camelot and the 1960s Epistemological Revolution: Rethinking the Politics–Patronage–Social Science Nexus," *Social Studies of Science* 31, no. 2 (2001): 171–206.

37. John L. Kennedy to Herbert Simon and proposal for "Experimental Research in the Behavioral Science and Regional Development" (May 26, 1955), HSP, Box 6, ff 210.

38. The following paraphrase is drawn from several dozen articles published in leading economic journals between 1948 and 1958. Three of the best exemplars are Arrow, *Social Choice and Individual Values*; Marschak, "Rational Behavior, Uncertain Prospects, and Measurable Utility."; and Paul Samuelson, "A Note on the Pure Theory of Consumer's Behaviour," *Economica* 5, no. 17 (1938): 61–71.

39. Arrow, *Social Choice and Individual Values*.

40. The impossibility theorem has been modified over time, but the original conditions were: *universality* (the ranking of social choices is unique and complete for everyone), *Pareto efficiency* (any different aggregation scheme would hurt someone's position for every person it helped), *non-dictatorship*, and the *independence of irrelevant alternatives* (changes in how one feels about C should not change how one ranks A and B).

41. See Anthony Downs, *An Economic Theory of Democracy* (New York: Harper & Brothers, 1957).

42. If one's preferences are transitive, then if one prefers A to B and B to C, then one will prefer A to C. This seems fairly reasonable, but there are some serious problems with this assumption: first, it assumes that preferences are context independent (i.e., that one does not prefer A to B on Monday and B to A on Tuesday and that A and B and C are all the same choices on Tuesday as they were on Monday); second, it assumes that A, B, and C are consistently identifiable units (like moves in a game); third, it assumes that people actually behave as if their preferences are transitive, when empirical studies have shown that they do not always do so.

43. Herbert A. Simon, "The Architecture of Complexity: Some Common Properties of Complex Systems" (Apr. 23, 1962), HSP, Box 2, ff 72.

44. *Well-defined* is a fascinating phrase that is often, ironically, rather poorly defined. It appears wherever people are attempting to define what can be programmed; in the language of this chapter, a *well-defined* situation is one in which all choices can be divided into unit choices and the set of possible unit choices is fully specified. The power—and the limits—of models of well-defined situations can be seen throughout the sciences of choice.

45. Robert M Thrall, Clyde Coombs, and Robert L. Davis, eds., *Decision Processes* (New York: Wiley, 1954).

46. Allen Newell and Herbert Simon, "Report on a General Problem-Solving Program" (Dec. 20, 1958), HSP, Box 12, ff: "UNESCO Paper: Report on a General Problem-Solving Program."

47. See Herbert A Simon, *The Sciences of the Artificial* (Cambridge, MA: MIT Press, 1969).

48. Ronald Coase, Martin Shubik, and others have pursued research on the economics of

information and communication. See Martin Shubik, "Information, Risk, Ignorance, and Indeterminacy," *Quarterly Journal of Economics* 68, no. 4 (1954): 629–40; Ronald Coase, "The New Institutional Economics," *AER* 88, no. 2 (1998): 72–74; Anoop Madhok, "Reassessing the Fundamentals and Beyond: Ronald Coase, the Transaction Cost and Resource-Based Theories of the Firm and the Institutional Structure of Production," *Strategic Management Journal* 23, no. 6 (2002): 535–50; Richard A. Posner, "Nobel Laureate: Ronald Coase and Methodology," *Journal of Economic Perspectives* 7, no. 4 (1993): 195–210; and Martin Shubik, "Information, Rationality, and Free Choice in a Future Democratic Society," *Daedalus* 96, no. 3 (1967): 771–78, and "Information, Theories of Competition, and the Theory of Games," *Journal of Political Economy* 60, no. 2 (1952): 145–50.

49. See Herbert A Simon, *The New Science of Management Decision* (Evanston, IL: Harper & Row, 1960).

50. Daniel Bell, *The Coming of Post-Industrial Society: A Venture in Social Forecasting*, new ed. (New York: Basic, 1999).

51. Allen Newell and Herbert A. Simon, *Human Problem Solving* (Englewood Cliffs, NJ: Prentice-Hall, 1972).

52. Carter A. Daniel, *MBA: The First Century* (Lewisburg, PA: Bucknell University Press, 1998); John A. Byrne, *The Whiz Kids: The Founding Fathers of American Business—and the Legacy They Left Us* (New York: Currency Doubleday, 1993); Robert Gleeson, "The Rise of Graduate Management Education in American Universities, 1908–1970" (PhD diss., Carnegie-Mellon University, 1997); Robert A. Gordon, and James E. Howell, *Higher Education for Business* (New York: Columbia University Press, 1959); Robert R. Locke, *Management and Higher Education since 1940: The Influence of America and Japan on West Germany, Great Britain, and France* (New York: Cambridge University Press, 1989).

53. For how narrow definitions of rationality in terms of immediate self-interest exclude much that looks perfectly rational from a broader perspective in the Prisoner's Dilemma, see Anatol Rapoport and Albert M Chammah, *Prisoner's Dilemma: A Study in Conflict and Cooperation* (Ann Arbor: University of Michigan Press, 1965).

54. Crowther-Heyck, *Herbert A. Simon*.

55. Samuelson, "Consumption Theory in Terms of Revealed Preference."

56. For a fascinating discussion of "outcomes" (such as purchases or votes) in social science, see Andrew Abbott, "The Idea of Outcome in US Sociology," in *The Politics of Method in the Human Sciences*, ed. George Steinmetz (Durham, NC: Duke University Press, 2005).

57. Allen Newell and Joseph Kruskal, "Organization Theory in Miniature" (May 18, 1951), HSP, Box 4, ff 147.

58. Vannevar Bush, "As We May Think," *The Atlantic*, July 1945, 101–8, 108.

59. Cowles Commission for Research in Economics, "Decision Making under Uncertainty: Eleventh Progress Report" (Mar. 1954), HSP, Box 6, ff 229, and "Fifth Report on the Project 'Theory of Resources Allocation,' under Subcontract with the RAND Corporation" (July 31, 1950), HSP, Box 6, ff 226; Clifford Hildreth, *The Cowles Commission in Chicago, 1939–1955* (New York: Springer-Verlag, 1986).

60. On the Behavioral Models Project at Columbia, see Angela M O'Rand, "Mathematizing Social Science in the 1950s: The Early Development and Diffusion of Game Theory," in *Toward a History of Game Theory: Supplement to the History of Political Economy*, ed. E. Roy Weintraub (Durham, NC: Duke University Press, 1992).

61. See Crowther-Heyck, *Herbert A. Simon*; Daniel, *MBA: The First Century*; and Robert Gleeson and Steven Schlossman, "George Leland Bach and the Rebirth of Graduate Management Education in the United States, 1945–1975," *Selections* 11, no. 3 (1995): 8–36.

62. This discussion draws in particular from Harvey, *Brief History of Neoliberalism*; Howard

Brick, *Transcending Capitalism: Visions of a New Society in Modern American Thought* (Ithaca, NY: Cornell University Press, 2006); and Amadae, *Rationalizing Capitalist Democracy*.

63. J. Cohen-Cole, "The Creative American: Cold War Salons, Social Science, and the Cure for Modern Society," *Isis* 100, no. 2 (2009): 219–62; Marga Vicedo, "Cold War Emotions: The War Over Human Nature," in Solovey and Cravens, *Cold War Social Science*.

64. See, for example, Herbert Simon's analyses of creative thinking in "Discovery, Invention, and Development: Human Creative Thinking," *Proceedings of the National Academy of Sciences of the United States of America* 80, no. 14 (1983): 4570–71. On ideas about creativity in cognitive science more generally, see Cohen-Cole, "Creative American."

CHAPTER 5: Modernity and Social Change in American Social Science

Epigraph. Cyril Black, *The Dynamics of Modernization: A Study in Comparative History* (New York: Harper & Row, 1966), 1.

1. Dorothy Ross, *The Origins of American Social Science* (Cambridge, New York: Cambridge University Press, 1991), 8.

2. Stein, quoted in David Harvey, *The Condition of Postmodernity: An Enquiry in the Origins of Cultural Change* (Cambridge, MA: Basil Blackwell, 1990), 17.

3. William Ogburn, "Stationary and Changing Societies," *American Journal of Sociology* 42, no. 1 (1936): 16–31, 16.

4. Peter Laslett, *The World We Have Lost*, University Paperbacks (London: Methuen, 1965).

5. Michael Latham, *Modernization as Ideology: American Social Science and "Nation Building" in the Kennedy Era* (Chapel Hill: University of North Carolina Press, 2000); Nils Gilman, *Mandarins of the Future: Modernization Theory in Cold War America* (Baltimore: Johns Hopkins University Press, 2003). An exception to the exclusive American focus is David C. Engerman, *Modernization from the Other Shore: American Intellectuals and the Romance of Russian Development* (Cambridge, MA: Harvard University Press, 2003). Also see David C. Engerman, Nils Gilman, and Michael Latham, eds., *Staging Growth: Modernization, Development, and the Global Cold War* (Amherst: University of Massachusetts Press, 2003).

6. Parsons's famed "pattern variables," while not discussed in this chapter, were widely accepted when they confirmed the longstanding ideas discussed above and treated very skeptically when they did not. See Talcott Parsons, *The Social System* (New York: Free Press, 1951), and Talcott Parsons and Edward Shils, eds., *Toward a General Theory of Action* (Cambridge, MA: Harvard University Press, 1951).

7. Michael Adas, *Machines as the Measure of Men: Science, Technology, and Ideologies of Western Dominance* (Ithaca, NY: Cornell University Press, 1989); Michael Adas, *Dominance by Design: Technological Imperatives and America's Civilizing Mission* (Cambridge, MA: Belknap Press of Harvard University Press, 2006).

8. Adas, *Machines as the Measure of Men*. Also see Lynnette Regouby, "The Limits of Nature: Alexander von Humboldt's Vision of Climate, Cultivation, and Culture in the Spanish Colonies" (MA thesis, University of Oklahoma, 2006).

9. Cyril Edwin Black, *Dynamics of Modernization: A Study in Comparative History* (New York: Harper & Row, 1966), 1–2. Walt W. Rostow, "The Stages of Economic Growth," *Economic History Review*, new series 12, no. 1 (1959): 1–16, 4.

10. Peter L. Berger, Brigitte Berger, and Hansfried Kellner, *The Homeless Mind: Modernization and Consciousness* (New York: Random House, 1973).

11. See William Ogburn, "Culture and Sociology," *Social Forces* 16 (1937): 161–69. Other relevant pieces by Ogburn include *Social Change with Respect to Culture and Original Nature* (London: G. Allen & Unwin, 1923); "Technology and Sociology," *Social Forces* 17 (1938): 1–8; and "Social Trends," *AJS* 45 (1940): 756–69. On Ogburn, technology, and social change, see

Rudi Volti, "Review: William F. Ogburn on 'Social Change with Respect to Culture and Origi-nal Nature,'" *Technology and Culture* 45 (2004): 396–405. On the history of the concept of cul-ture, see Susan Hegeman, *Patterns for America: Modernism and the Concept of Culture* (Princeton, NJ: Princeton University Press, 1999); John S. Gilkeson, "The Domestication of 'Culture' in Interwar America, 1919–1941," in *The Estate of Social Knowledge*, JoAnne Brown and David Van Keuren, eds. (Baltimore, MD: Johns Hopkins University Press, 1991); John S. Gilkeson, *Anthro-pologists and the Rediscovery of America, 1886–1965* (New York: Cambridge University Press, 2010); Glen Gendzel, "Political Culture: Genealogy of a Concept," *Journal of Interdisciplinary History* 28, no. 2 (1997): 225–50; and George W. Stocking, *Race, Culture, and Evolution: Essays in the History of Anthropology* (Chicago: University of Chicago Press, 1982).

12. Rostow, "Stages of Economic Growth," 15.

13. Ibid., 14.

14. Black, *Dynamics of Modernization*, 7.

15. See Dorothy Ross, "Grand Narrative in American Historical Writing: From Romance to Uncertainty," *American Historical Review* 100, no. 3 (1995): 651–77.

16. On the importance of the distinction between these visions of the role of the social scientist in a democracy, see Hunter Crowther-Heyck, *Herbert A. Simon: The Bounds of Reason in Modern America* (Baltimore: Johns Hopkins University Press, 2005), esp. chapters 2–4.

17. Rostow, "Stages of Economic Growth," 11–12.

18. Peter J. Bowler, *Evolution: The History of an Idea*, 3d ed. (Berkeley: University of Califor-nia Press, 2003).

19. Ogburn, "Culture and Sociology," 162. On Weismann's work as a significant dividing line in the social interpretation of evolution, and on the centrality of Malthusianism to such debates, see Piers J. Hale, *Political Descent: Malthus, Mutualism, and the Politics of Evolution in Victorian England* (Chicago: University of Chicago Press, 2014).

20. Karl Deutsch, for example, saw the analysis of nation-building as a kind of exercise in cybernetic social physiology in *The Nerves of Government: Models of Political Communication and Control* (New York: Free Press, 1963).

21. John C. Caldwell and Bruce Caldwell, *Demographic Transition Theory* (Dordrecht: Springer, 2006); Frank Trovato, *Population and Society: Essential Readings* (Don Mills, ON: Ox-ford University Press, 2002).

22. See Thorstein Veblen, "Is Economics an Evolutionary Science?" in *The Place of Science in Modern Civilization and Other Essays* (New Brunswick, NJ: Transaction, 1990).

23. Walt Rostow, "Review: Toward a General Theory of Action," *World Politics* 5, no. 4 (1953): 530–54.

24. William Fielding Ogburn, *Social Change with Respect to Culture and Original Nature*, rev. ed. (New York: Viking, 1950).

25. Black, *Dynamics of Modernization*, 7.

26. Herbert Butterfield, *The Origins of Modern Science, 1300–1800* (London: G. Bell, 1949). On postwar reinterpretations of the scientific revolution, see Hunter Crowther-Heyck, "A. R. Hall's Scientific Revolution," *H-Ideas Retrospective Reviews*, Feb. 2001, www.h-net.org/reviews/showrev.php?id=4938.

27. Rostow, "Stages of Economic Growth," 4.

28. Robert K. Merton, *Science, Technology, and Society in Seventeenth Century England* (New York: Howard Fertig, 1993).

29. This close connection to policy and its effects on modernization theory's prominence is discussed in Engerman, Latham, and Gilman, *Staging Growth*. For a related analysis of the effects of shifts in political culture upon social research, see Mark Solovey, "Project Camelot and the 1960s Epistemological Revolution: Rethinking the Politics-Patronage-Social Science Nexus," *So-*

cial Studies of Science 31, no. 2 (2001): 171–206, and *Shaky Foundations: The Politics-Patronage-Social Science Nexus in Cold War America* (New Brunswick, NJ: Rutgers University Press, 2013).

30. Paul Samuelson, "Economics in a Golden Age," in *Paul Samuelson and Modern Economic Theory*, ed. E. Cary Brown and Robert M. Solow (New York: McGraw-Hill, 1983).

31. Hunter Crowther-Heyck, "Patrons of the Revolution: Ideals and Institutions in Postwar Social Science," *Isis* 97, no. 3 (2006): 420–46; Walter Heller, *New Dimensions of Political Economy* (Cambridge, MA: Harvard University Press, 1966); Assembly of Behavioral and Social Sciences, Study Project on Social Research and Development, *The Federal Investment in Knowledge of Social Problems* (Washington, DC: National Academy of Sciences, 1978); Mary Furner and Barry Supple, eds., *The State and Economic Knowledge: The American and British Experience* (Washington, DC: Woodrow Wilson International Center for Scholars and Cambridge University Press, 1990); Michael Lacey and Mary Furner, eds., *The State and Social Investigation in Britain and the United States* (Washington, DC: Woodrow Wilson International Center for Scholars and Cambridge University Press, 1993); Michael Bernstein, *A Perilous Progress: Economists and Public Purpose in Twentieth-Century America* (Princeton, NJ: Princeton University Press, 2001); Sarah Igo, *The Averaged American: Surveys, Citizens, and the Making of a Mass Public* (Cambridge, MA: Harvard University Press, 2007); Alice O'Connor, *Poverty Knowledge: Social Science, Social Policy, and the Poor in Twentieth-Century US History* (Princeton, NJ: Princeton University Press).

32. Contrast Fukuyama's exuberance in *The End of History and the Last Man* (New York: Free Press, 1992) with David Harvey's less-than-cheerful *Condition of Postmodernity* (Cambridge, MA: Basil Blackwell, 1990), for example.

33. Mark Solovey, "Riding Natural Scientists' Coattails onto the Endless Frontier: The SSRC and the Quest for Scientific Legitimacy," *Journal of the History of the Behavioral Sciences* 40, no. 4 (2004): 393–422.

34. Latham, *Modernization as Ideology*.

35. Manuel Castells, *The Rise of the Network Society* (Malden, MA: Blackwell Publishers, 1996).

36. Fukuyama, *End of History*.

37. Daniel Bell, *The Coming of Post-Industrial Society: A Venture in Social Forecasting*, new ed. (New York: Basic, 1999).

38. Ibid., lxxxiii. Bell writes, "I think of society as comprising three different realms that hang together over time in different ways and that move in different historical rhythms. These realms are the techno-economic *system*, the political *order*, and the cultural *sphere*" (lxxxii). Emphasis in original.

39. William H. McNeill, *The Rise of the West: A History of the Human Community* (Chicago: University of Chicago Press, 1963); Paul M. Kennedy, *The Rise and Fall of the Great Powers: Economic Change and Military Conflict from 1500 to 2000* (New York: Random House, 1987); Samuel P. Huntington, *The Clash of Civilizations and the Remaking of World Order* (New York: Simon & Schuster, 1996); Edmund Burke and Kenneth Pomeranz, *The Environment and World History* (Berkeley: University of California Press, 2009); Bruno Latour, *We Have Never Been Modern* (Cambridge, MA: Harvard University Press, 1993).

CHAPTER 6: A Model Science?

Epigraph. C. M. Arensberg, "American Communities," *AA*, 57, no. 6 (1955): 1143–62.

1. Robert M. Solow, "How Did Economics Get That Way and What Way Did It Get?" *Daedalus* 126, no. 1 (1997): 39–58, 42, 43, 57.

2. In this chapter, information from the main survey (see chapter 1, appendix) is supplemented by a second survey, focused exclusively on usage of the word *model*. This second survey included the five journals in the first survey—*AA*, *AER*, *AJS*, *APSR*, and *PR*—and added

Philosophy of Science as it includes philosophical discussion of models and theories. This second survey employed a different methodology than the first, in part due to the advent of some new tools in the JSTOR and PyschLit databases that made it possible to search for keywords and have the keywords appear in "context" (a small amount of text before and after the keyword, usually enough to determine the sense in which the term was being used). The search methods returned slightly different sets of articles for the five social science journals in the overlapping years (1925–1975), and I did not perform the same evaluative reading of every article for this second survey, so the data sets are not directly comparable. Hence, results have not been merged into the same tables, and I distinguish between conclusions drawn from the first, evaluative survey, versus the second, "keyword-in-context" survey. The data in this paragraph comes from the second survey.

3. Nancy Cartwright, *The Dappled World: A Study of the Boundaries of Science* (Cambridge: Cambridge University Press, 1999).

4. One central aspect of the "semantic view" of models (in simplified form) is that models are the central unit of scientific "theorizing"; in general, adherents of the semantic view *also* see models as *set-theoretic structures* that are either isomorphic with or strongly similar in structure to the "target system" being represented. Bas Van Fraassen is one of the best known exemplars of this approach; see *Scientific Representation: Paradoxes of Perspective* (Oxford: Oxford University Press, 2010). For other perspectives on models, see Daniela Bailer-Jones, *Scientific Models in Philosophy of Science* (Pittsburgh: University of Pittsburgh Press, 2009).

5. Roman Frigg and Stephan Hartmann, "Models in Science," *The Stanford Encyclopedia of Philosophy*, Fall 2012 ed., ed. Edward N. Zalta, http://plato.stanford.edu/archives/fall2012/entries/models-science/.

6. This idea of "manipulable mobiles" is indebted, equal parts, to Bruno Latour's discussion of "immutable mobiles" in "Visualization and Cognition: Thinking with Eyes and Hands," *Knowledge and Society* 6, no. 1 (1986): 1–40, and to Mary Morgan's discussion of models in economics as manipulable representations of economic ideas in *The World in the Model: How Economists Work and Think* (New York: Cambridge University Press, 2012). Melvin Kranzberg's dictum that "technologies come in packages, large and small" also played a role, for models, in this perspective, look like packaged "intellectual technologies" (a term coined by Daniel Bell long ago). Melvin Kranzberg, "Technology and History: 'Kranzberg's Laws,' " *Technology and Culture* 27 (1986): 544–60; Daniel Bell, *The Coming of Post-Industrial Society: A Venture in Social Forecasting*, special anniversary ed. (New York: Basic Books, 1999), xciv.

7. Joel Isaac, *Working Knowledge: Making the Human Sciences from Parsons to Kuhn* (Cambridge, MA: Harvard University Press, 2012). Isaac's arguments here build on those advanced in "Tool Shock: Technique and Epistemology in the Postwar Social Sciences," *History of Political Economy* 42, suppl. 1 (2010): 133–64.

8. Bailer-Jones, *Scientific Models in Philosophy of Science*.

9. Frigg and Hartmann, "Models in Science."

10. Bailer-Jones, *Scientific Models in Philosophy of Science*, 1.

11. Margaret Morrison, "Models as Autonomous Agents," in Mary S. Morgan and Margaret Morrison, eds., *Models as Mediators: Perspectives on Natural and Social Science* (Cambridge, UK: Cambridge University Press, 2000): 38–65, 38.

12. Eric Winsberg, "Simulations, Models, and Theories: Complex Physical Systems and Their Representations," *Philosophy of Science* 68, no. 3 (2001): S442–54.

13. Lorenzo Magnani and Nancy J. Nersessian, *Model-Based Reasoning: Science, Technology, Values* (New York: Kluwer Academic, 2002).

14. Max Black, *Models and Metaphors; Studies in Language and Philosophy* (Ithaca, NY: Cornell University Press, 1962).

15. Alan D. Schrift, *The History of Continental Philosophy* (Chicago: University of Chicago Press, 2010).

16. Frigg and Hartmann, "Models in Science."

17. Morgan, *World in the Model*, 2.

18. Ibid., 2. Morgan follows Alistair Crombie's notion of scientific styles very closely. See A. C. Crombie, *Styles of Scientific Thinking in the European Tradition: The History of Argument and Explanation Especially in the Mathematical and Biomedical Sciences and Arts*, 3 vols. (London: Duckworth, 1994).

19. Morgan, *World in the Model*, 5.

20. Ibid., 20.

21. Latour, "Visualization and Cognition," 20–22.

22. See also Lorraine Daston and Peter Galison, "The Image of Objectivity," *Representations* 40 (Fall 1992): 81–128.

23. And if they can be rescaled, as they can in Latour's theory, then one version of such an immutable mobile can be reduced to another, which would seem to violate his philosophical principle of nonreduction as laid out in the second half of *The Pasteurization of France* (Cambridge, MA: Harvard University Press, 1988).

24. Morgan, *World in the Model*, 31. Emphasis in original.

25. Ibid., 34.

26. Jennifer Karns Alexander, *The Mantra of Efficiency: From Waterwheel to Social Control* (Baltimore: Johns Hopkins University Press, 2008).

27. See, for example, Mary B. Hesse, *Models and Analogies in Science* (Notre Dame, IN: University of Notre Dame Press, 1966), and May Brodbeck, *Readings in the Philosophy of the Social Sciences* (New York: Macmillan, 1968).

28. See, for example, Allen Newell, J. C. Shaw, and Herbert A. Simon, "Elements of a Theory of Human Problem Solving," *Psychological Review* 65, no. 3 (1958): 151–66, 151.

29. Malcolm Bradbury and James Walter McFarlane, *Modernism: 1890–1930* (Harmondsworth, UK: Penguin, 1976).

30. Bradbury and McFarlane, *Modernism*, 21, 25.

31. Ross, *Origins of American Social Science*.

32. Bradbury and McFarlane, *Modernism*, 50.

33. Ibid., 48.

34. Thomas Haskell, *The Emergence of Professional Social Science: The ASSA and the Nineteenth Century Crisis of Authority* (Urbana: University of Illinois Press, 1977).

35. Theodore Porter, *Trust in Numbers: The Pursuit of Objectivity in Science and Public Life* (Princeton, NJ: Princeton University Press, 1995).

36. Lisa Gitelman and Geoffrey B. Pingree, *New Media, 1740–1915* (Cambridge, MA: MIT Press, 2003).

37. Bernard Lightman, "The Visual Theology of Victorian Popularizers of Science. From Reverent Eye to Chemical Retina," *Isis* 91, no. 4 (2000): 651–80.

38. Peter Louis Galison discusses the different epistemological status of pictures versus counts in physics in *Image and Logic: A Material Culture of Microphysics* (Chicago: University of Chicago Press, 1997).

39. Stephen Kern, *The Culture of Time and Space, 1880–1918* (Cambridge, MA: Harvard University Press, 2003).

40. Arturo Rosenblueth, Norbert Wiener, and Julian Bigelow, "Behavior, Purpose, and Teleology," *Philosophy of Science* 10, no. 1 (1943): 18–24.

41. Arturo Rosenblueth and Norbert Wiener, "The Role of Models in Science," *Philosophy of Science* 12, no. 4 (1945): 316–21.

42. Ibid., 316. While they admired the goals of operationalist philosophers and scientists, such as Percy Bridgman, Wiener and Rosenblueth wrote, "Not all scientific questions are directly amenable to experiment. There is a hierarchy of questions whose levels are determined by the generality of the answers sought" (ibid.).

43. Ibid., 317.

44. Ibid., 320, 321.

45. F. S. C. Northrop, "The Neurological and Behavioristic Psychological Basis of the Ordering of Society by Means of Ideas," *Science* 107, no. 2782 (1948): 411–17, 412.

46. Ibid.

47. Ibid., 416.

48. K. W. Deutsch, "Mechanism, Organism, and Society: Some Models in Natural and Social Science," *Philosophy of Science* 18, no. 3 (1951): 230–52, 230.

49. Ibid., 230; emphasis added. This recalls Herbert Simon's statement that "manipulating symbols is as concrete as sawing boards in a carpentry shop." Herbert A. Simon, *Models of My Life* (New York: Basic Books, 1991), 193.

50. Carl H. Denbow, "Is Mathematics a Formal Discipline?" *Philosophy of Science* 22, no. 2 (1955): 161–64, 161.

51. C. West Churchman, *Introduction to Operations Research* (New York: Wiley, 1957), 71, 72, 110, 71, 72.

52. Ibid., 155, 157.

53. Ibid., 157.

54. Herbert A. Simon, *Models of Man: Social and Rational. Mathematical Essays on Rational Behavior in a Social Setting* (New York: Wiley, 1957), 199.

55. Anthony F. C. Wallace, "Revitalization Movements," *AA* 58, no. 2 (1956): 264–81, 266.

56. J.C.R. Licklider and Robert Taylor, "The Computer as a Communications Device," *Science and Technology* 76 (1968): 21–31, 29.

57. Daniel T. Rodgers, *Age of Fracture* (Cambridge, MA: Belknap Press of Harvard University Press, 2011). Also see Ben Orlove, "From Culture to Knowledge: Climate, Sustainable Development, and 'Traditional Environmental Knowledge,'" (presented at the University of Oklahoma, Feb. 15, 2006).

58. Hunter Heyck, "Embodiment, Emotion, and Moral Experiences: The Human and the Machine in Film," in *Science Fiction and Computing: Essays on Interlinked Domains*, ed. David L. Ferro and Eric G. Swedin (New York: McFarland, 2011), 230–48. Fom the mid-nineteenth century to the 1990s, one could not get a patent for a "business method" that was not materialized as a piece of machinery, with rare exceptions (mostly having to do with financial practices). Over the course of the 1980s and 1990s, companies applied for thousands of patents on the grounds that they were materializing new business methods in new machinery by doing things on a digital computer or doing them online. The patent office slowly came around to accepting this view, and in 1998–99 it began handing out business model patents like they were free samples at a legal convention. This caused an uproar, and the patent office and the Supreme Court have been trying to figure out a reasonable middle ground ever since. Doing so has proven to be quite difficult.

59. This abstraction of the idea of the machine, it should be noted, is likely related to the rapid rise in popular usage of the term *model* in the 1990s and 2000s, when suddenly news pages were filled with discussions not only of business models and economic models but also of fee-for-service models, the Canadian model, the Japanese model, the corporate model, and so on. In all of these cases, this looser popular usage was not of models as *manipulable mobiles*, for though such models did refer to simplified representations, they did not refer to rule-bound ones or to ones that satisfied Latour's criteria for mobility. Rather, these representations

were hypersimplified, extraordinarily abstracted reductions to a basic mechanism or function. As such, they represented machines so idealized and abstracted they had to be virtual.

60. http://www.youtube.com/watch?v=owI7DOeO_yg.

CONCLUSION: History and Legacy, the Tree and the Web

Epigraph. Jerome Bruner, Jacqueline Goodnow, and George Austin, *A Study of Thinking* (New York: Wiley, 1956), 2.

1. Some examples include Fred M. Kaplan, *The Wizards of Armageddon,* Stanford Nuclear Age Series (Stanford, CA: Stanford University Press, 1991); James A. Smith, *The Idea Brokers: Think Tanks and the Rise of the New Policy Elite* (New York: Free Press, 1991); Michael Lacey and Mary Furner, eds., *The State and Social Investigation in Britain and the United States* (Washington, DC: Woodrow Wilson International Center for Scholars and Cambridge University Press, 1993); Robert M. Collins, *More: The Politics of Economic Growth in Postwar America* (New York: Oxford University Press, 2000); Michael E. Latham, *Modernization as Ideology: American Social Science and "Nation Building" in the Kennedy Era,* The New Cold War History (Chapel Hill: University of North Carolina Press, 2000); Michael Bernstein, *A Perilous Progress: Economists and Public Purpose in Twentieth-Century America* (Princeton, NJ: Princeton University Press, 2001); Alice O'Connor, *Poverty Knowledge: Social Science, Social Policy, and the Poor in Twentieth-Century US History, Politics and Society in Twentieth-Century America* (Princeton, NJ: Princeton University Press, 2001); Jennifer S. Light, *From Warfare to Welfare: Defense Intellectuals and Urban Problems in Cold War America* (Baltimore: Johns Hopkins University Press, 2003); Mark Solovey and Hamilton Cravens, *Cold War Social Science: Knowledge Production, Liberal Democracy, and Human Nature,* 1st ed. (New York: Palgrave Macmillan, 2012).

2. Barry D. Karl, *Charles E. Merriam and the Study of Politics* (Chicago: University of Chicago Press, 1974); James Kloppenberg, *Uncertain Victory: Social Democracy and Progressivism in European and American Thought, 1870–1920* (New York: Oxford University Press, 1986); Rivka Shpak Lissak, *Pluralism and Progressives: Hull House and the New Immigrants, 1890–1919* (Chicago: University of Chicago Press, 1989); Mark C. Smith, *Social Science in the Crucible: The American Debate over Objectivity and Purpose, 1918–1941* (Durham, NC: Duke University Press, 1994), Martin Bulmer, *The Chicago School of Sociology: Institutionalization, Diversity, and the Rise of Sociological Research,* ed. Morris Janowitz, Heritage of Sociology Series (Chicago: University of Chicago Press, 1984); Donald Fisher, *Fundamental Development of the Social Sciences: Rockefeller Philanthropy and the United States Social Science Research Council* (Ann Arbor: University of Michigan Press, 1993); Kurt Danziger, *Constructing the Subject: Historical Origins of Psychological Research* (New York: Cambridge University Press, 1990); Robert N. Proctor, *Value-Free Science? Purity and Power in Modern Knowledge* (Cambridge, MA: Harvard University Press, 1991); Katherine Pandora, *Rebels within the Ranks: Psychologists' Critique of Scientific Authority and Democratic Realities in New Deal America,* Cambridge Studies in the History of Psychology (Cambridge: Cambridge University Press, 1997); Edward A. Purcell Jr., *The Crisis of Democratic Theory: Scientific Naturalism and the Problem of Value* (Lexington: University Press of Kentucky, 1973).

3. These interests were in keeping with the broader fascination with professionalization during the 1970s–80s, with law and medicine receiving the most attention. Some exemplary works include Burton Bledstein, *The Culture of Professionalism: The Middle Class and the Development of Higher Education in America* (New York: W. W. Norton, 1976); Alexandra Oleson and John Voss, eds., *The Organization of Knowledge in Modern America, 1860–1920* (Baltimore: Johns Hopkins University Press, 1979); Paul Starr, *The Social Transformation of American Medicine* (New York: Basic Books, 1982); Andrew Abbott, *The System of Professions: An Essay on the Division of Expert Labor* (Chicago: University of Chicago Press, 1988); and Brian Balogh, "Reorganizing the Organizational Synthesis: Federal-Professional Relations in Modern America," *Studies*

in American Political Development 5 (Spring 1991): 119–72. A point of transition from the earlier professionalization literature to a more constructivist/cultural-authority approach is Theodore Porter, *Trust in Numbers: The Pursuit of Objectivity in Science and Public Life* (Princeton, NJ: Princeton University Press, 1995), which combined interest in professionalization and disciplinary identity with a new look at rhetorical strategies (such as quantification) and their role in the construction of cultural authority.

4. Alvin W. Gouldner, *The Coming Crisis of Western Sociology* (New York: Basic Books, 1970); Raymond Seidelman, *Disenchanted Realists: Political Science and the American Crisis, 1884–1984* (Albany: State University of New York Press, 1985); Jeffrey C. Alexander, *Twenty Lectures: Sociological Theory since World War II* (New York: Columbia University Press, 1987); James Farr and Richard Seidelman, eds., *Discipline and History: Political Science in the United States* (Ann Arbor: University of Michigan Press, 1993); John Gunnell, *The Descent of Political Theory: The Genealogy of an American Vocation* (Chicago: University of Chicago Press, 1993).

5. Dorothy Ross, *The Origins of American Social Science* (New York: Cambridge University Press, 1991); Mary Furner, *Advocacy and Objectivity: A Crisis in the Professionalization of American Social Science, 1865–1905* (Lexington: University Press of Kentucky for the OAH, 1975); Robert C Bannister, *Sociology and Scientism: The American Quest for Objectivity, 1880–1940* (Chapel Hill: University of North Carolina Press, 1987); Thomas Haskell, *The Emergence of Professional Social Science: The ASSA and the Nineteenth-Century Crisis of Authority* (Urbana: University of Illinois Press, 1977).

6. Paul Edwards, *The Closed World: Computers and the Politics of Discourse in Cold War America* (Cambridge, MA: MIT Press, 1996); N. Katherine Hayles, *How We Became Posthuman: Virtual Bodies in Cybernetics, Literature, and Informatics* (Chicago: University of Chicago Press, 1999); Philip Mirowski, *Machine Dreams: Economics Becomes a Cyborg Science* (New York: Cambridge University Press, 2002); Rebecca M. Lemov, *World as Laboratory: Experiments with Mice, Mazes, and Men* (New York: Hill & Wang, 2005); Hunter Crowther-Heyck, "Patrons of the Revolution: Ideals and Institutions in Postwar Social Science," *Isis* 97, no. 3 (2006): 420–46; Sarah Elizabeth Igo, *The Averaged American: Surveys, Citizens, and the Making of a Mass Public* (Cambridge, MA: Harvard University Press, 2007); Jamie Cohen-Cole, "The Creative American: Cold War Salons, Social Science, and the Cure for Modern Society," *Isis* 100, no. 2 (2009): 219–62; David C. Engerman, *Know Your Enemy: The Rise and Fall of America's Soviet Experts* (New York: Oxford University Press, 2009).

7. By the mid-1990s, even work on Progressive Era social science was beginning to look different, with a much heavier emphasis on practices and applications (as opposed to conceptual schemes) for the cultural authority of social science: see, for example, Elizabeth Lunbeck, *The Psychiatric Persuasion: Knowledge, Gender, and Power in Modern America* (Princeton, NJ: Princeton University Press, 1994), and Daniel T. Rodgers, *Atlantic Crossings: Social Politics in a Progressive Age* (Cambridge, MA: Belknap Press of Harvard University Press, 1998).

8. I am using the word *technoscience* here in a different way than would Bruno Latour, the person who has done the most to define and popularize the term. My usage here is narrower than his in that I primarily am using it to refer to an institutionalized fusion of science and technology on the levels of ideas, practices, patrons, and goals, typically sponsored by states or corporate entities with significant resources. Latour would remove the latter constraint regarding sponsors and add that technoscience is pursued in or constituted by a network of actors whose relationships are structured in particular ways: e.g., by the existence of "obligatory points of passage" and "centers of calculation," and the use of "immutable mobiles" to represent the world in ways that enable some actors to enroll others as allies. Latour's concept of technoscience emphasizes the ever-presence of an inner politics within scientific and technical pursuits; the contestants in this inner politics may draw from resources available to them

through participation in other, perhaps larger, networks, each with its own politics. Doing so then links the networks, changing the flows of power and knowledge within them. There is much of great value in Latour's schema, though it is so broad that it seems to encompass all institutionalized scientific and technical pursuits, making it less useful as a term for describing some of the novel features of postwar science, medicine, and engineering. One could say, then, that my *technoscience* is a specific instantiation of Latour's universal *technoscience*, one that is largely peculiar to the postwar world. See Bruno Latour, *Science in Action, How to Follow Scientists and Engineers through Society* (Cambridge, MA: Harvard University Press, 1987).

On "technopolitics," see Gabrielle Hecht, *The Radiance of France: Nuclear Power and National Identity after World War II* (Cambridge, MA: MIT Press, 1998). In addition to Latour, on "technoscience," see Don Ihde and Evan Selinger, eds., *Chasing Technoscience: Matrix for Materiality* (Bloomington: Indiana University Press, 2003); Thomas Parke Hughes, Agatha C. Hughes, Michael Thad Allen, and Gabrielle Hecht, eds., *Technologies of Power: Essays in Honor of Thomas Parke Hughes and Agatha Chipley Hughes* (Cambridge, MA: MIT Press, 2001); and John Law, *Aircraft Stories: Decentering the Object in Technoscience* (Durham, NC: Duke University Press, 2002); and, though he does not use the term, see also Peter Louis Galison, *Image and Logic: A Material Culture of Microphysics* (Chicago: University of Chicago Press, 1997).

9. On infrastructural projects as technosocial projects, see Paul N. Edwards, "Meteorology as Infrastructural Globalism," *Osiris* 21(2006): 229–50; Thomas J. Misa, Philip Brey, and Andrew Feenberg, *Modernity and Technology* (Cambridge, MA: MIT Press, 2003).

10. For missile defense, see the many works that discuss SAGE, such as Thomas Parke Hughes, *Rescuing Prometheus* (New York: Pantheon, 1998). On the related issue of the aerospace industry and its related technosocial projects, see Steven Johnson, "Three Approaches to Big Technology: Operations Research, Systems Engineering, and Project Management," *Technology and Culture* 38, no. 4 (1997): 891–919; Walter McDougall, *The Heavens and the Earth: A Political History of the Space Program* (New York: Basic Books, 1985); P. Redfield, "The Half-Life of Empire in Outer Space," *Social Studies of Science* 32, no. 5/6 (2002): 791–826; and Slava Gerovitch, " 'New Soviet Man' Inside Machine: Human Engineering, Spacecraft Design, and the Construction of Communism," *Osiris* 22 (2007): 135–57. For planned cities, see Jennifer S. Light, *The Nature of Cities: Ecological Visions and the American Urban Professions, 1920–1960* (Baltimore: Johns Hopkins University Press, 2009) and *From Warfare to Welfare: Defense Intellectuals and Urban Problems in Cold War America* (Baltimore: Johns Hopkins University Press, 2003); Lloyd Rodwin, *Planning Urban Growth and Regional Development: The Experience of the Guayana Program of Venezuela* (Cambridge, MA: MIT Press, 1969); Stuart W. Leslie and Robert Kargon, "Exporting MIT: Science, Technology, and Nation-Building in India and Iran," *Osiris* 21 (2006): 110–30; James Holston, *The Modernist City: An Anthropological Critique of Brasília* (Chicago: University of Chicago Press, 1989); and Felix Driver and David Gilbert, *Imperial Cities: Landscape, Display, and Identity* (Manchester, UK: Manchester University Press, 1999). For surveillance systems, see Kristie Macrakis, *Seduced by Secrets: Inside the Stasi's Spy-Tech World* (New York: Cambridge University Press, 2008), and Reginald Whitaker, *The End of Privacy: How Total Surveillance Is Becoming a Reality* (New York: New Press, 1999). There is an enormous literature on development projects and policy as technosocial endeavors; for a superb example, see Suzanne Moon, "Justice, Geography, and Steel: Technology and National Identity in Indonesian Industrialization," *Osiris* 24 (2009): 253–77.

11. John Law, "Technology and Heterogeneous Engineering: The Case of Portuguese Expansion," in *The Social Construction of Technological Systems*, ed. Wiebe Bijker, Thomas P. Hughes, and Trevor Pinch (Cambridge, MA: MIT Press, 1987), 111–34.

12. Langdon Winner, *The Whale and the Reactor: A Search for Limits in an Age of High Technology* (Chicago: University of Chicago Press, 1986).

13. Bijker, Hughes, and Pinch, *Social Construction of Technological Systems*; Thomas Parke Hughes, *Networks of Power: Electrification in Western Society, 1880–1930* (Baltimore: Johns Hopkins University Press, 1983).

14. Dean E Wooldridge, *Mechanical Man: The Physical Basis of Intelligent Life* (New York: McGraw-Hill, 1968), *The Machinery of Life* (New York: McGraw-Hill, 1966), and *The Machinery of the Brain* (New York: McGraw-Hill, 1963).

15. Will Thomas, "Beyond the Science of Warfare: The Origins of Systems Analysis, 1937–1954" (presented at the History of Science Society Annual Meeting, Arlington, VA, November 2007); Andrew Pickering, "Cybernetics and the Mangle: Ashby, Beer, and Pask," *Social Studies of Science* 32, no. 3 (2002): 413–37; Mirowski, *Machine Dreams*; Edwards, *Closed World*; Donna Haraway, "A Cyborg Manifesto: Science, Technology, and Socialist-Feminism in the Late Twentieth Century," in *Simians, Cyborgs, and Women: The Reinvention of Nature*, ed. Donna Haraway (London: Free Association, 1991), 149–82. One way of tracking the influence of such projects would be to trace the spread of flow charts the way David Kaiser follows Feynman diagrams in *Drawing Theories Apart: The Dispersion of Feynman Diagrams in Postwar Physics* (Chicago: University of Chicago Press, 2005).

16. Paul Edwards, *The Closed World: Computers and the Politics of Discourse in Cold War America* (Cambridge, MA: MIT Press, 1996).

17. On the reformation of the self as a key part of twentieth-century human science, see Greg Eghigian, Andreas Killen, and Christine Leuenberger, "Introduction: The Self as Project: Politics and the Human Science in the Twentieth Century," *Osiris* 22 (2007): 1–25.

18. Daniel T. Rodgers, *Age of Fracture* (Cambridge, MA: Belknap Press of Harvard University Press, 2011).

19. Samuelson puts forth his economic theory in its most sophisticated form in *Foundations of Economic Analysis* (Cambridge, MA: Harvard University Press, 1947).

20. Paul Samuelson, *Economics: An Introductory Analysis*, 1st ed. (New York: McGraw-Hill, 1948), 412.

21. Manuel Castells, *The Rise of the Network Society* (Malden, MA: Blackwell Publishers, 1996).

22. Charles Murray, *Losing Ground: American Social Policy, 1950–1980* (New York: Basic Books, 1984).

Index